T0226493

DIOXIN-CONTAINING WASTES

DIOXIN-CONTAINING WASTES

WASTES

Treatment Technologies

by

M. Arienti, L. Wilk,
M. Jasinski, N. Prominski

Alliance Technologies Corporation
Bedford, Massachusetts

NOYES DATA CORPORATION
Park Ridge, New Jersey, U.S.A.

Copyright © 1988 by Noyes Data Corporation
Library of Congress Catalog Card Number 88-17858
ISBN: 0-8155-1181-7
ISSN: 0090-516X
Printed and bound in the United Kingdom

Transferred to Digital Printing, 2011

Library of Congress Cataloging-in-Publication Data

Dioxin-containing wastes : treatment technologies / by M. Arienti . . .
   [et al.].
         p. cm. -- (Pollution technology review, ISSN 0090-516X ; no.
   160)
      Bibliography: p.
      Includes index.
      ISBN 0-8155-1181-7 :
      1. Dioxins--Environmental aspects. 2. Chemical industries--Waste
   disposal--Environmental aspects. 3. Hazardous wastes--Environmental
   aspects. I. Arienti, M. II. Series.
   TD196.C5D553   1988
   628.5'46--dc19   88-17858
                        CIP

Foreword

This book provides descriptions of treatment technologies for dioxin-containing wastes. It covers proven technology on waste management options and should be a useful tool for those involved in decision making on dioxin-containing wastes. Information is included on processes that have been evaluated with actual dioxin waste streams, and processes that have been tested using similar waste streams. In addition to process-specific details, extensive data are included on the characterization of dioxin wastes in general.

The 1984 Hazardous and Solid Waste Act Amendments to the Resource Conservation and Recovery Act (RCRA) directed EPA to ban certain dioxin-containing wastes from land disposal, unless EPA determines that restrictions on land disposal of these wastes are not needed to protect human health and the environment. An important aspect of the land disposal restrictions is the identification and evaluation of alternative technologies that can be used to treat the listed wastes in such a way as to meet proposed treatment levels which EPA has determined are protective of human health and the environment. The purpose of this book is to identify and evaluate the alternative technologies that remove and/or destroy dioxin and related compounds from listed dioxin wastes in order to achieve constituent levels that allow the safe land disposal of the treated residues.

Technologies evaluated are those that destroy or somehow change the form of dioxin so that it is less toxic. The majority of the technologies described are those whose performance has been tested on dioxin-containing wastes. Those that have not been tested on dioxin-containing wastes have, at least, been tested on PCB-containing wastes. Because of the similarity of PCBs and dioxins, these technologies should also be applicable to dioxin wastes. Technologies that have been developed to full scale as well as those only investigated in the laboratory are included.

The information in the book is from *Technical Resource Document: Treatment Technologies for Dioxin-Containing Wastes,* prepared by M. Arienti, L. Wilk, M. Jasinski, and N. Prominski of Alliance Technologies Corporation for the U.S. Environmental Protection Agency, October 1986.

The table of contents is organized in such a way as to serve as a subject index and provides easy access to the information contained in the book.

Advanced composition and production methods developed by Noyes Data Corporation are employed to bring this durably bound book to you in a minimum of time. Special techniques are used to close the gap between "manuscript" and "completed book." In order to keep the price of the book to a reasonable level, it has been partially reproduced by photo-offset directly from the original report and the cost saving passed on to the reader. Due to this method of publishing, certain portions of the book may be less legible than desired.

NOTICE

The materials in this book were prepared as accounts of work sponsored by the U.S. Environmental Protection Agency. It has been subject to the Agency's peer and administrative review and has been approved for publication. On this basis the Publisher assumes no responsibility nor liability for errors or any consequences arising from the use of the information contained herein.

Mention of trade names or commercial products or processes does not constitute endorsement or recommendation for use by the Agency or the Publisher. Final determination of the suitability of any information or product for use contemplated by any user, and the manner of that use, is the sole responsibility of the user.

The book is intended for informational purposes only. The reader is warned that caution must always be exercised when dealing with hazardous chemicals such as dioxins, hazardous wastes containing dioxins, or hazardous processes involving dioxins; and expert advice should be obtained before implementation is considered.

Contents and Subject Index

1. Executive Summary

INTRODUCTION

The 1984 Hazardous and Solid Waste Act Amendments to the Resource
Conservation and Recovery Act (RCRA) directed EPA to ban certain
dioxin-containing wastes from land disposal unless EPA determines that
restrictions on land disposal of these wastes are not needed to protect human
health and the environment. Congress, through the 1984 Amendments, fixed a
deadline of 24 months from the enactment of the Amendments for EPA to regulate
the land disposal of these identified wastes (with some exceptions). In the
event that the Agency has not issued regulations by that time (November 1986),
land disposal of all specified dioxin-containing waste streams automatically
will be banned.

An important aspect of the land disposal restrictions is the
identification and evaluation of alternative technologies that can be used to
treat the listed wastes in such a way as to meet proposed treatment levels
which EPA has determined are protective of human health and the environment.
If alternatives to land disposal are not available by November 1986, it may be
necessary to extend the deadline for the restrictions on land disposal. The
purpose of this document is to identify and evaluate alternative technologies
that remove and/or destroy dioxin and related compounds from listed dioxin
wastes in order to achieve constituent levels that allow the safe land
disposal of the treated residues.

SCOPE

A number of potential technologies exist for treating wastes containing
dioxin. Because many of the technologies are currently in early stages of
development, it is not possible to fully assess the effectiveness of these

technologies at this time. Further testing of a technology in the future may, for example, indicate that a technology is or is not practical on a full scale. In addition, several new technologies for treating dioxin wastes may emerge for which information is not currently available. Consequently, it must be emphasized that the information discussed here represents that which was available in the spring of 1986.

Technologies under evaluation are those that destroy or somehow change the form of dioxin so that it is less toxic. Temporary management methods, such as "storage in mines," are not evaluated because these methods only involve moving the waste without changing the chemical form and characteristics of the waste. The majority of the technologies are those whose performance has been tested on dioxin-containing wastes. Those that have not been tested on dioxin-containing wastes have, at least, been tested on PCB-containing wastes. Because of the similarity of PCBs and dioxins, these technologies should also be applicable to dioxin wastes. Technologies that have been developed to full scale as well as those only investigated in the laboratory are included. This is primarily because, as mentioned previously, this field is rapidly evolving. Many of the technologies that are now only in the laboratory stage may be standard technologies for treatment of these wastes in the future.

DEFINITION OF DIOXIN WASTE

The term "dioxin waste" is meant to include those RCRA wastes listed as EPA hazardous waste Numbers F021, F022, F023, F026 and F027. As shown in Table 1.1, these waste codes are designated as "acute hazardous" and include wastes from the production and manufacturing use of tri-, tetra-, and pentachlorophenols, wastes from the manufacturing use of tetra-, penta-, and hexachlorobenzene under alkaline conditions, and also discarded, unused formulations containing tri- tetra-, and pentachlorophenols. Soil that has been contaminated by improper management of these wastes is also encompassed by these waste codes. Residues from the incineration of this contaminated soil are designated as toxic instead of acute hazardous and are covered under waste code F028.

TABLE 1.1.   DIOXIN CONTAMINATED WASTES LISTED AS RCRA HAZARDOUS WASTES,
JANUARY 14, 1985, 50 FR 1978

Hazardous Waste from Nonspecific Source

EPA hazardous waste no.	Hazardous waste	Hazard code
F020*	Wastes** from the production or manufacturing use of tri- or tetrachlorophenol, or of intermediates used to produce their derivatives.**	(H)
F021*	Wastes** from the production or manufacturing use of pentachlorophenol (PCP), or of intermediates used to produce its derivatives.	(H)
F022*	Wastes** from the manufacturing use of tetra-, penta-, or hexachlorobenzene under alkaline conditions.	(H)
F023*	Wastes** from the production of materials on equipment previously used for the production or manufacturing use of tri- or tetrachlorophenols.***	(H)
F026*	Wastes** from the production of materials on equipment previously used for the manufacturing of tetra-, penta-, or hexachlorobenzene under alkaline conditions.*	(H)
F027*	Discarded unused formulations containing tri-, tetra-, or pentachlorophenol or discarded unused formulations derived from these chlorophenols.****	(H)
F028	Residues resulting from the incineration or thermal treatment of soil contaminated with EPA hazardous waste F020, F021, F022, F023, F026, and F027.	(T)

*A proposed regulation [50 FR 37338] would make residues from the incineration of these wastes (if the waste contained less than or equal to 10 ppm TCDD prior to incineration) toxic instead of acute hazardous .

**Except wastewater and spent carbon from hydrogen chloride purification.

***This listing does not include wastes from the production of hexachlorophene from highly purified 2,4,5-trichlorophenol.

****This listing does not include formulations containing hexachlorophene synthesized from prepurified 2,4,5-trichlorophenol as the sole component.

(H) = Acute Hazardous Waste

(T) = Toxic Waste

The wastes described by these waste codes are listed hazardous wastes primarily because they contain one of a number of forms of dioxin. The term "dioxin" is one which has been used very loosely. Dioxin encompasses a family of aromatic compounds known chemically as dibenzo-p-dioxin. The forms of dioxin that are of most environmental concern are the chlorinated dioxins, in which a chlorine atom occupies one or more of the available eight positions on the double benzene ring structure. Thus, there are 75 possible chlorinated dioxin compounds, the most toxic of which is 2,3,7,8-tetrachlorodibenzo-p-dioxin (TCDD). Throughout this report, various terms will be used to refer to certain types of dioxin. When only the word "dioxin" is used, it refers to chlorinated dioxin compounds in general. Other commonly used abbreviations are listed below:

PCDDs   = all isomers of chlorinated dibenzo-p-dioxins

CDDs   = all isomers of tetra-, penta-, and hexachlorodibenzo-p-dioxins

TCDD   = the 2,3,7,8- isomer

PeCDD,
HxCDD,   = the penta-, hexa-, and octachloro compounds
and OCDD

Other toxic constituents that may be present in the listed dioxin wastes are chlorinated dibenzofurans (CDFs), chlorophenols, and chlorophenoxy compounds.

WASTE SOURCES, CHARACTERISTICS, AND QUANTITIES

The waste codes included in the dioxin listing encompass process wastes from the production of various chlorophenols, primarily 2,4,5-trichlorophenol and pentachlorophenol, and chlorophenoxy pesticides such as 2,4,5-T and Silvex. As indicated in a report prepared by Technical Resources, Inc. for the EPA Office of Solid Waste, the manufacture of most of these compounds has been stopped. For example, 2,4,5-trichlorophenol has not been manufactured for several years. As a result, the majority of the dioxin-bearing process wastes requiring treatment at this time are wastes such as still bottoms and reactor residues that were generated in the past and remain to be treated. The only process waste stream that is still being generated, and may continue to be generated in the future, is from the manufacture of pentachlorophenol (PCP). However, by far the largest quantity of dioxin-bearing wastes that have been identified are the contaminated soils such as those at Times Beach, Missouri, and various other CERCLA sites throughout the country.

Table 1.2 shows estimated waste quantities for each of the waste codes. Several items associated with the information in the table should be noted. One is that no sources have yet been identified for waste codes F022 and F026. Another is that waste code F028 is not included because it is expected that residues from future incineration of contaminated soil will meet EPA delisting requirements. Finally, contaminated soils are placed in a separate category both because of their unique physical form relative to most process wastes, and also because a large fraction of the contaminated soils are at CERCLA sites whose wastes will not be affected by the RCRA land disposal restrictions until November 1988.

The estimates of the quantities of wastes generated within each waste category in Table 1.2 could have a significant impact on future treatment practices. As shown in the table, there are more than 500,000 metric tons (MT) of dioxin-contaminated soil that may require treatment. This quantity is considerably greater than the estimated maximum 7500 MT of process wastes, such as still bottoms currently requiring treatment and the estimated 2500 MT of industrial process wastes that will be generated in future years. Consequently, it would appear that treatment technologies capable of treating soil wastes are of most importance at this time, particularly those technologies, such as solvent extraction, that are capable of removing the toxic constituents from the soil and thereby reducing the total volume of waste requiring final detoxification/destruction.

TECHNOLOGIES FOR TREATING DIOXIN WASTES

As mentioned previously, a number of technologies for treating dioxin waste are evaluated in this document. A summary of the status of these technologies is provided in Table 1.3. Because studies have shown that dioxin

TABLE 1.2.   SUMMARY OF DIOXIN WASTE SOURCES AND QUANTITIES

Waste code	Waste source	Physical form	Quantity generated (metric tons) Present (or stored)	Future
F020	Manufacture of herbicides such as 2,4,5-T, 2,4,5-trichlorophenol, hexachlorophene; disposal of wastes in uncontrolled landfills or storage areas	- Still bottoms containing organic solvents and chlorophenols - Nonaqueous phase leachate (NAPL) containing solvents, chlorophenols, heavy metals - Carbon used to treat aqueous leachate	Still bottoms - 2,300 NAPL - 1,450 Other - 550	0 0 - 200 Unknown
F021	Manufacture of pentachlorophenol: wastes from purification; wastes from formulation	- Still bottoms or other concentrated materials containing nonvolatile organic solids and chlorinated solvents and phenols - Sludges from formulation	Still bottoms - 0 Formulation waste-700	750 Unknown
F022	No known sources at this time	- NA*	0	0
F023	Production of chemicals on equipment formerly used to manufacture F020 compounds, e.g., 2,4-D on 2,4,5-T equipment	- Similar to F020 wastes - still bottoms, reactor residues containing chlorophenols and organic solvents, and wash water sludges from formulation	0 - 600	0 - 600
F026	No known sources at this time	- NA	0	0
F027	Discarded formulation of tri-, tetra-, and pentachlorophenols and their derivatives	- Active ingredient in an emulsifiable concentrate, as a salt or an ester, or dissolved in an oil (such as in the case of pentachlorophenol	1000 - 2000	0-1,000***
	Contaminated soil from improper disposal and spills of F020-F027**	- Soils containing low concentrations of dioxins and related compounds	500,000	Unknown

*NA - Not applicable.
**Not listed as a specific waste code
***Only from pentachlorophenol products

TABLE 1.3. SUMMARY OF TREATMENT PROCESSES

Process name	Applicable waste streams	Stage of development	Performance/ destruction achieved	Cost	Residuals generated
Stationary Rotary Kiln Incineration	Solids, liquids, sludges	Several approved and commercially available units for PCBs; not yet used for dioxins	Greater than six nines DRE for PCBs; greater than five nines DRE demonstrated on dioxin at combustion research facility	$0.25 - $0.70/lb for PCB solids	Treated waste material (ash), scrubber wastewater, particulate from air filters, gaseous products of combustion
Mobile Rotary Kiln Incineration	Solids, liquids, sludges	EPA mobile unit is permitted to treat dioxin wastes; ENSCO unit has been demonstrated on PCB waste	Greater than six nines DRE for dioxin by EPA unit; process residuals delisted	NA*	Same as above.
Liquid Injection Incineration	Liquids or sludges with viscosity less than 10,000 ssu (i.e., pumpable)	Full scale land-based units permitted for PCBs; only ocean incinerators have handled dioxin wastes	Greater than six nines DRE on PCB wastes; ocean incinerators only demonstrated three nines on dioxin containing herbicide orange	$200 - $500/ton	Same as above, but ash is usually minor because solid feeds are not treated
Fluidized-bed Incineration (Circulating Bed Combustor)	Solids, sludges	GA Technologies mobile circulating bed combustor has a TSCA permit to burn PCBs anywhere in the nation; not tested yet on dioxin	Greater than six nines DRE demonstrated by GA unit on PCBs	$60 - $320/ton for GA unit	Treated waste (ash), parti- culates from air filters
High Temperature Fluid Wall (Huber AER)	Primarily for granular contaminated soils, but may also handle liquids	Huber stationary unit is permitted to do research on dioxin wastes; pilot scale mobile reactor has been tested at several locations on dioxin contaminated soils	Pilot scale mobile unit demonstrated greater than five nines DRE on TCDD - contaminated soil at Times Beach (79 ppb reduced to below detection)	$300 - $600/ton	Treated waste solids (converted to glass beads), particulates from baghouse, gaseous effluent (primarily nitrogen)
Infrared Incinerator (Shirco)	Contaminated soils/sludges	Pilot scale, port- able unit tested on waste containing dioxin; full scale units have been used in other appli- cations; not yet permitted for TCDD	Greater than six nines DRE on TCDD-contaminated soil	Treatment costs are $200 - $1,200 per ton	Treated material (ash); particulates captured by scrubber (separated from scrubber water)

(continued)

*Not available

TABLE 1.3 (continued)

Process name	Applicable waste streams	Stage of development	Performance/ destruction achieved	Cost	Residuals generated
Molten Salt (Rockwell Unit)	Solids, liquids, sludges; high ash content wastes may be troublesome	Pilot scale unit was tested on various wastes - further development is not known	Up to eleven nines DRE on hexachlorobenzene; greater than six nines DRE on PCB using bench scale reactor	NA	Spent molten salt containing ash, particulates from baghouse
Supercritical Water Oxidation	Aqueous solutions or slurries with less than 20 percent organics can be handled	Pilot scale unit tested on dioxin-containing wastes - results not yet published	Six nines DRE on dioxin-containing waste reported by developer, but not presented in literature; lab testing showed greater than 99.99% conversion of organic chloride for wastes containing PCB	$0.32 - $2.00/ gallon, $77 - $480/ton	High purity water, inorganic salts, carbon dioxide, nitrogen
Plasma Arc Pyrolysis	Liquid waste streams (possibly low viscosity sludges)	Prototype unit (same as full scale) currently being field tested	Greater than six nines destruction of PCBs and CCl_4	$300 - $1,400/ton	Exhaust gases (H_2 and CO) which are flared and scrubber water containing particulates
In Situ Vitrification	Contaminated soil - soil type is not expected to affect the process	Full scale on radioactive waste; pilot scale on organic contaminated wastes	Greater than 99.9% destruction efficiency (DE) (not offgas treatment system) on PCB-contaminated soil	$120 - $250/$m^3$	Stable/immobile molten glass; volatile organic combustion products (collected and treated)
Solvent Extraction	Soils, still bottoms	Full scale still bottoms extraction has been tested - pilot scale soils washer needs further investigation	Still bottom extraction: 340 ppm TCDD reduced to 0.2 ppm; 60-90% removal from soils, but reduction to below 1 ppb not achieved	NA	Treated waste material (soil, organic liquid); solvent extract with concentrated TCDD
Stabilization/ Fixation	Contaminated soil	Laboratory scale using cement and emulsified asphalt; lab tests also using K-20	Tests using cement showed decreased leaching of TCDD, but up to 27% loss of stabilized material due to weathering followed by leaching	NA	Stabilized matrix (soil plus cement, asphalt, or other stabilization material); matrix will still contain TCDD

*Not available

(continued)

TABLE 1.3 (continued)

Process name	Applicable waste streams	Stage of development	Performance/ destruction achieved	Cost	Residuals generated
UV Photolysis	Liquids, still bottoms, and soils can be treated if dioxin is first extracted or desorbed into liquid	Full scale solvent extraction/UV process was used to treat 4,300 gallons of still bottoms in 1980; thermal desorption/UV process currently undergoing second field test	Greater than 98.7% reduction of TCDD using solvent extraction/UV process - residuals contained ppm concentrations of TCDD; thermal desorption/UV process demonstrated reduction of TCDD in soil to below 1 ppb	Cost of treating the 4,300 gallons of still bottoms using solvent extraction/UV was $1 million; thermal desorption/UV estimated to cost $250 - $1,250/ton	Solvent extraction/UV process generated treated still bottoms, a solvent extract stream, and an aqueous salt stream; thermal desorption/UV generates a treated soil stream and a solvent extract stream
Chemical Dechlorination- APEG Processes	Contaminated soil (other variations of the process used to treat PCB-contaminated oils)	Slurry process currently being field tested at pilot scale; in situ process has been tested in the field	Laboratory research has demonstrated reduction of 2,000 ppb TCDD to below 1 ppb for slurry (batch process); laboratory and field testing of in situ process not as promising	$296/ton for in situ APEG process; $91/ton for slurry (batch) process	Treated soil containing chloride salts (reagent is recovered in the slurry process)
Biological Degradation- primarily in situ addition of microbes	Research has been directed toward in situ treatment of contaminated soils - liquids are also possible	Currently laboratory scale-field testing in next year or two	50-60% metabolism of 2,3,7,8-TCDD in a week long period under lab conditions using white rot fungus - reduction to below 1 ppb not achieved	NA	Treated waste medium such as soil or water with TCDD metabolites depending on microorganisms
Chemical Degradation using Ruthenium Tetroxide	Liquid or soil wastes - possible most effective in decontaminating furniture, other surfaces	Laboratory scale - no work reported since 1983	Reduction of 70 ppb TCDD to below 10 ppb in 1 hr (on soil sample)	NA	Treated medium plus the solvent which has been added (water, CCl₄); TCDD and products not known
Chemical Degradation using Chloroiodides	Liquid or soil - thought to be most applicable to decontaminating furniture and buildings	Laboratory scale - no work reported since 1983	Up to 92% degradation of TCDD in benzene - reduction to below 1 ppb were not demonstrated	NA	Treated waste medium; degradation products are chlorophenols
Gamma Ray Radiolysis	Liquid waste streams (has been applied to sewage sludge disinfection)	Laboratory research; no research currently being conducted	97% destruction of 2,3,7,8-TCDD in ethanol after 30 hours - 100 ppb to 3 ppb	Cost for sewage disinfection facility treating 4 tons per day is $40 per ton; TCDD treatment would be more expensive	Less chlorinated dioxin molecules are the degradation end products in addition to the treated waste medium

*Not available

decomposes by heating or oxidation at temperatures greater than 1000°C, thermal methods for treating these wastes have received a large amount of attention. Thermal technologies evaluated in this document are those in which heat is the major agent of treatment or destruction. Technologies inlcuded in this category are:

- Stationary rotary kiln incineration

- Mobile rotary kiln incineration

- Liquid injection incineration

- Fluidized-bed incineration

- Infrared incineration

- High temperature fluid wall destruction

- Plasma arc pyrolysis

- Molten salt destruction

- In situ vitrification

- Supercritical water oxidation

EPA has indicated that incineration is currently the only sufficiently demonstrated treatment technology for dioxin-containing waste (51 FR 1733). RCRA performance standards for incineration and other thermal treatment processes require the demonstration of 99.9999 percent destruction and removal efficiency (DRE) of the principal organic hazardous constituent (POHC). Several of the thermal technologies have demonstrated this performance on chlorinated compounds of one type or another. However, only three, and perhaps four, thermal technologies have been demonstrated to achieve this level of performance on dioxin. These technologies are the EPA mobile rotary kiln incinerator, Huber's high temperature fluid wall reactor, Shirco's infrared incinerator, and possibly, Modar's supercritical water oxidation process. Modar has not yet released data conclusively showing six nines DRE, but they do claim to have achieved this performance. Thermal technologies that have achieved six nines DRE on PCBs include stationary rotary kiln incinerators, liquid injection incinerators, fluidized-bed incinerators (the

circulating bed variation), the plasma arc process, and the molten salt
process. The in situ vitrification process has not shown six nines DRE;
however, it is as much a stabilization process as it is a destruction
process. Therefore, the primary objective of this technology is to prevent
the leaching of dioxin or other toxic constituents from the treated soil;
whether the dioxin is driven out of the soil by volatilization or merely
contained within the vitrified material is a secondary concern (as long as
volatilized dioxin is captured and subsequently destroyed).

Nonthermal technologies evaluated include the following:

- Chemical dechlorination

- Ultraviolet (UV) photolysis

- Solvent extraction

- Biodegradation

- Stabilization/fixation

- Chemical degradation using ruthenium tetroxide

- Chemical degradation using chloroiodides

- Gamma ray radiolysis

Of the nonthermal technologies, those that have shown the most promise
and the highest level of recent investigation and testing are chemical
dechlorination and UV photolysis. Both of these technologies are currently
being field tested on dioxin-contaminated soil. As indicated in Table 1.3,
preliminary field data on the thermal desorption/UV photolysis process
indicate that dioxin was desorbed from soil to a level below 1 ppb, and then
destroyed efficiently using ultraviolet radiation. The chemical
dechlorination process has also demonstrated a reduction of TCDD in soil to
below 1 ppb, but only on a laboratory scale.

The other nonthermal processes have not shown as much promise with regard
to treating dioxin waste. Solvent extraction is a potentially useful
technology since it could, if successfully applied to soil treatment, reduce
the volume of the waste stream that requires final treatment/destruction by
several orders of magnitude. Unfortunately, this technology has not yet

demonstrated the ability to reduce dioxin in contaminated soil to a level of 1 ppb. Biodegradation is also a potentially attractive approach since it presumably would not require the large energy inputs, sophisticated equipment, and the chemical additions that the other technologies require. However, biodegradation, particularly in situ, has not proven to be very effective as a dioxin destruction process. Stabilization and/or fixation would allow the treatment of contaminated soils in place. Since this method does not involve destruction of the dioxin there is always the possibility that the stabilized waste/soil matrix will break down and the dioxin will be released. Finally, the last three technologies listed (two chemical degradation processes and gamma ray radiolysis) are methods that have been studied in the laboratory but have not yet shown enough promise technically or economically to be developed on a larger scale. Investigation of these methods, at this time, appears to have stopped.

Of all the treatment technologies evaluated, none is currently commercially available for the treatment of dioxin wastes. The EPA mobile incinerator has been used to treat a variety of waste forms at the Denney Farm in Missouri, but this unit is intended to be used for research purposes and not as a commercial treatment process. The high temperature fluid wall process (AER) operated by Huber at its Borger, Texas facility is permitted to perform research on dioxin contaminated wastes and is also a research tool which is not intended to be used for actual waste treatment.

CONCLUSIONS

Dioxin wastes, particularly those dioxin-contaminated soils which account for over 98 percent of the contaminated wastes identified in Table 1.2, contain low levels (10 to 100 ppb) of dioxins and/or dibenzofurans. Nonetheless, many technologies, particularly the thermal destruction technologies, require that the total quantity of the waste be treated to destroy the extremely low dioxin fraction resulting in very high energy usage for dioxin destruction. In addition, when incineration and other thermal destruction technologies are used, large quantities of exhaust gases are generally formed. These waste streams can contain toxic products of incomplete combustion (PICs) and other hazardous emissions. They and other

associated waste streams are themselves subject to costly treatment
processes. Therefore, technologies such as solvent extraction or desorption,
which separate the toxic constituents from the waste matrix prior to final
treatment should receive further investigation.

Most of the emerging technologies are being designed for operation at the
waste source. This trend to portable or field-erected technologies reflects a
reaction to public opposition to the transport of dioxin waste from source to
waste treatment facilities, and should continue to be encouraged.

In addition, because of the large volume of soil contaminated by
relatively low concentrations of dioxin, it is also important to investigate
methods of in situ treatment. These methods would limit the handling of the
waste so that further dispersion of contaminated materials into the
environment is minimized. Most of the technologies in this category, such as
biodegradation, in situ vitrification, chemical dechlorination, and
stabilization in the near future have not yet been sufficiently demonstrated.
Use in the near future seems improbable without more intense development of
these technologies. Steps should be taken to encourage these developments.

The treatment of dioxin contaminated liquids and low viscosity sludges
does not appear to be as large a problem as is the treatment of contaminated
soils. This is primarily because the quantity of liquids and sludges is much
lower, and also because the liquid waste form generally calls for less
extensive handling and pretreatment. Technologies, such as plasma arc
pyrolysis and supercritical water oxidation, appear to be capable of treating
these wastes, and their development should be fostered, as should other
reasonable activities aimed at the development of emerging technologies.

2. Regulations Concerning Management of Listed Dioxin Wastes

2.1 CURRENT REGULATION UNDER RCRA

Certain dioxin contaminated wastes originally regulated under the Toxic Substance Control Act (TSCA), 40 CFR Part 775, were listed as hazardous wastes under the Resource Conservation and Recovery Act (RCRA) on January 14, 1985, 50 FR 1978. The January 14, 1985 RCRA Amendments list as acute hazardous wastes certain chlorinated dibenzo-p-dioxins, dibenzofurans, and phenols (and their phenoxy derivatives). A complete listing was presented in Table 1.1. When the RCRA Amendment listing dioxin-contaminated wastes became effective on July 15, 1985, duplicate listings of certain dioxins under RCRA and TSCA were revoked.

The inclusion of these dioxin-contaminated wastes under the RCRA regulations was mandated by the RCRA statutory amendments entitled the Hazardous and Solid Waste Amendments of 1984 (HSWA), signed into law November 8, 1984 as Public Law 98-616. HSWA, among other things, mandate a RCRA listing status for 2,3,7,8-tetrachlorodibenzo-p-dioxin (TCDD)-contaminated wastes, stringent technical requirements for land disposal facilities, an expanded definition of land disposal, and various land disposal bans and restrictions.

HSWA state that in the case of any hazardous waste which is prohibited from one or more methods of land disposal, the storage of such hazardous waste is prohibited unless such storage is solely for the purpose of the accumulation of such quantities of hazardous waste as are necessary to facilitate its proper recovery, treatment, or disposal. In order for interim storage of these wastes to be excluded from this prohibition, it must be demonstrated that the storage is solely for the purposes of facilitating proper recovery, treatment, or disposal. HSWA also specify a two year (24 months) period during which EPA must decide whether or not to completely

14

prohibit the landfilling of "dioxin-contaminated wastes." However, there is a four year exemption for response actions taken under CERCLA Section 104 or 106. Therefore, the land disposal of untreated (or unstabilized) dioxin-contaminated materials will not be prohibited for CERCLA actions until November 9, 1988.

HSWA also state that storage (now included within the definition of land disposal) of hazardous wastes in mines or caves is specifically prohibited until such time as a permit has been issued under RCRA Section 3005(c). In addition, placement of noncontainerized, bulk solids or bulk liquid waste in underground mines or caves is prohibited until the EPA Administrator has determined that such placement is protective of human health. In addition,, HSWA specify more stringent minimum technology requirements for all new land disposal facilities.

The RCRA regulatory amendments of January 14, 1985 (effective July 9, 1985), listing certain dioxin-contaminated wastes as RCRA acute hazardous wastes, include specific provisions for the management of these wastes. These acute hazardous wastes are subject to the 1 kg small quantity generator limitation, and residues in empty containers will also be regulated. Residues from the incineration of dioxin-contaminated soils are listed only as toxic RCRA wastes and are not as stringently controlled. In addition, a proposed rule in 50 FR 37338 would also make residues from the incineration of other dioxin-containing wastes (F020-F027) toxic instead of acute hazardous wastes, if the wastes contained less than 10 ppm TCDD prior to incineration. However, this proposed rule will most likely not be finalized before the fall of 1986. All persons who generate, transport, treat, store, or dispose of the listed wastes were required to notify EPA or an authorized State by April 15, 1985. All hazardous waste management facilities which treat, store, or dispose of the listed wastes, and which qualify to handle the listed wastes under interim status, were required to notify by April 15, 1985 and submit a Part A permit application by July 15, 1985. Even those sites that had qualified for interim status prior to the regulatory amendments are not allowed to handle the listed wastes unless they qualify to handle these wastes, notify EPA or an authorized state by April 15, 1985, and submit a Part A application by July 15, 1985.

Pertinent regulatory provisions are summarized below.

- 2,3,7,8-TCDD-contaminated wastes resulting from the production or manufacturing use of several chlorophenols and chlorobenzenes, including contaminated soil, are added to the list of RCRA regulated acute hazardous wastes (RCRA hazardous waste numbers F020, F021, F022, F023, F026, F027, F028); RCRA regulated quantity for small quantity generators is 1 kilogram of 2,3,7,8-TCDD-contaminated material;

- 2,3,7,8-TCDD wastes may be disposed only in fully permitted RCRA (Part B) land disposal facilities (interim status land disposal facilities are not acceptable);

- Interim status facilities that may be acceptable for the management of 2,3,7,8-TCDD wastes include surface impoundments (for wastewater sludge; managed pursuant to 40 CFR 264.231), enclosed waste piles (pursuant to 40 CFR 264.250(c), tanks (pursuant to 40 CFR 264.200), containers (pursuant to 40 CFR 264.175), incinerators (if certified pursuant to 265.352), and thermal treatment units (if certified pursuant to 265.383);

- A waste management plan is required for all land disposal facilities that submit Part B of their RCRA permit application. The waste management plan will specifically address the means by which the waste will be managed safely at the land disposal facility;

- 2,3,7,8-TCDD wastes may not be stored or disposed of in unlined units;

- Interim status incinerators and interim status thermal treatment units are allowed to burn 2,3,7,8-TCDD wastes if they are "certified" by the Assistant Administrator for the EPA Office of Solid Waste and Emergency Response, pursuant to 40 CFR 265.352 and .383, respectively as meeting 40 CFR Part 264, Subpart O, RCRA performance standards;

- Incinerators and thermal treatment units that are used to burn 2,3,7,8-TCDD wastes must achieve a DRE of 99.9999 percent (i.e., six nines DRE); and

- Residue resulting from incineration or thermal treatment of dioxin-containing soils (F028) must be, at a minimum, managed at a RCRA interim status land disposal facility.

2.2   PROPOSED LAND DISPOSAL RESTRICTIONS

On January 14, 1986 EPA proposed a framework for a regulatory program to
implement the congressionally mandated land disposal prohibition
[52 FR 1602]. The regulatory framework establishes a new 40 CFR Part 268
which specifies treatment standards and effective dates for the
dioxin-contaminated wastes and prohibits land disposal of these wastes unless
the specified treatment standards are achieved. EPA has proposed specific
constituent screening levels in 40 CFR 268.42 for the dioxin-contaminated
wastes, F020, F021, F022, F023, F026, and F027. Waste No. F028 is the residue
from incineration or thermal treatment of dioxin-contaminated soil to
six nines DRE. Waste No. F028 is considered toxic (as opposed to acute
hazardous) and therefore, is not addressed by EPA in these proposed RCRA
regulatory amendments.

The proposed treatment standards are based on specific constituent
screening levels. These screening levels, shown in Table 2.1, are determined
through a toxicity characteristic leaching procedure (detailed in Appendix I
of 51 FR 1750) which tests the extract of the waste for the concentration of
constituents of concern. The EPA procedure for analyzing the extracts
(Method 8280) has a detection limit of 1.0 ppb for CDDS and CDFs. Therefore,
at this time, the land disposal restrictions specify that the residuals from
treatment of listed dioxin wastes must contain less than 1.0 ppb of
extractable CDD and CDFs in order for them to be land disposed as nonhazardous
materials. As detection limits are lowered, the treatment standards will
approach the constituent screening levels. Nonetheless, all land disposal
must meet RCRA regulations established in the January 14, 1985 regulations.

The proposed regulations specify certain prohibitions on storage.
Section 268.50 states that dioxin-contaminated hazardous waste not meeting
specified treatment standards may not be stored in tanks or containers after
November 8, 1986, unless:

1.   The owner or operator of a hazardous waste treatment, storage, or
      disposal facility stores such waste for 90 days or less; or

2.   A transporter stores manifested shipments of such waste in
      containers at a transfer facility for 10 days or less; or

TABLE 2.1.   CONCENTRATIONS OF CONSTITUENTS OF CONCERN WHICH WILL RESULT
IN BANNING LISTED WASTES FROM LAND DISPOSAL [51 FR 1732]

Constituent*	Screening level (mg/l)
2,3,7,8-TCDD	4×10^{-9}
Other TCDDs	4×10^{-7}
2,3,7,8-PeCDDs	8×10^{-9}
Other PeCDDs	8×10^{-7}
2,3,7,8-HxCDDs	1×10^{-7}
Other HxCDDs	1×10^{-5}
2,3,7,8-TCDFs	4×10^{-8}
Other TCDFs	4×10^{-6}
2,3,7,8-PeCDFs	4×10^{-8}
Other PeCDFs	4×10^{-8}
2,3,7,8-HxCDFs	4×10^{-7}
Other HxCDFs	4×10^{-5}
2,4,5-Trichlorophenol	8.0
2,4,6,-Trichlorophenol	0.04
2,3,4,6-Tetrachlorophenol	2.0
Pentachlorophenol	1.0

*Definitions of abbreviations used above

TCDDs and TCDFs = All isomers of tetrachlordibenzo-p-dioxins and
-dibenzofurans respectively.

PeCDDs and PeCDFs = The pentachlorodibenzo-p-dioxins and -dibenzofurans.

HxCDDs and HxCDFs = The hexa-isomers.

3.   Such waste is accumulated onsite by the operator and does not exceed
the applicable time limitations set forth in 40 CFR 262.34.

The proposed regulations also provide the following variances to the land
disposal prohibition effective November 8, 1986:

1.   Extensions past the effective date of the land disposal prohibition
may be obtained from EPA on a case-by-case basis if the applicant
demonstrates that, among other things, an alternative technology
will protect human health and the environment and that this
alternate technology will not be available by the effective date
and/or capacity for existing technologies will not be available by
the effective date, however, all efforts have been made to meet the
deadline (40 CFR 268.4).

2.   EPA will consider petitions to allow land disposal in particular
limits past the effective date of the land disposal prohibition if
the applicant demonstrates that, among other things, there will be
no adverse impact on human health and the environment and no
migration of hazardous constituents (40 CFR 268.5).

3.   All wastes meeting the treatment levels specified in 40 CFR 268.42
may be land disposed in accordance with applicable RCRA regulations
after November 8, 1986.

4.   In the preamble to the proposed regulations, EPA states that a
nationwide variance of up to 2 years may be granted if alternate
recovery and disposal capacity is inadequate nationwide.

3. Characterization and Quantification of Listed Dioxin Wastes

3.1 INTRODUCTION

The purpose of this section is to characterize the wastes described by
RCRA codes F020, F021, F022, F023, F026, and F027. These codes descriDe
wastes from the production and manufacturing use of tri-, tetra-, and
pentachlorophenols and from the manufacturing use of tetra-, penta-, and
hexachlorobenzenes under alkaline conditions and elevated temperatures. These
wastes include still bottoms, reactor residues, untreated brines, spent filter
aids, spent carbon adsorbents, and sludges resulting from wastewater
treatment. They also include wastes resulting from the production of
materials on equipment previously used for the production and manufacturing of
tri- and tetrachlorophenols, and formulations containing these chlorophenols
and their derivatives. Waste code F028 is a treatment residue from
incineration or thermal treatment of dioxin-containing soil to six nines DRE.
It is designated a toxic and not an acute hazardous waste, and therefore is
not addressed in this document. The untreated soils, however, that have been
contaminated by spills of wastes in codes F020, F021, F022, F023, F026 and
F027 are defined as hazardous (50 FR 28713) and are addressed.

As shown in Table 3.1, the basis for listing each of these wastes and for
banning them from land disposal is the expected or known presence of
significant quantities of tetra-, penta-, and hexachlorodibenzo-p-dioxins
(CDDs) and chlorinated dibenzofurans (CDFs). These compounds are among the
most potent animal carcinogens known and are potential human carcinogens in
addition to being extremely persistent in the environment. These wastes may
also contain significant concentrations of tri-, tetra-, and
pentachlorophenols and their chlorophenoxy derivatives, some of which are
potential human carcinogens (2,4,6-TCP) and/or are suspected of causing liver
and kidney damage (U.S. EPA, 1985).

TABLE 3.1. BASIS FOR LISTING WASTES [50 FR 1978]

EPA code	Waste code description	Hazardous constituents for which listed
F020	Wastes (except wastewater and spent carbon from hydrogen chloride purification) from the production or manufacturing use (as a reactant, chemical intermediate, or other component in a formulating process) of tri- or tetrachlorophenol, or of intermediates used to produce their pesticide derivatives. (This listing does not include wastes from the production of Hexachlorophene from highly purified 2,4,5-trichlorophenol).	Tetra- and pentachlorodibenzo-p-dioxins; tetra and pentachlorodibenzofurans; tri- and tetrachlorophenols and their chlorophenoxy derivative acids, esters, ethers, amines and salts.
F021	Wastes (except wastewater and spent carbon from hydrogen chloride purification) from the production or manufacturing use (as a reactant, chemical intermediate, or component in a formulating process) of pentachlorophenol, or of intermediates used to produce its derivatives.	Penta- and hexachlorodibenzo-p-dioxins; penta- and hexachlorodibenofurans; pentachlorophenol and its derivatives.
F022	Wastes (except wastewater and spent carbon from hydrogen chloride purification) from the manufacturing use (as a reactant, chemical intermediate, or component in a formulating process) of tetra-, penta-, or hexachlorobenzenes under alkaline conditions.	Tetra-, penta-, and hexachlorodibenzo-p-dioxins; tetra-, penta-, and hexachlorodibenzofurans.

(continued)

TABLE 3.1 (continued)

EPA code	Waste code description	Hazardous constituents for which listed
F023	Wastes (except wastewater and spent carbon from hydrogen chloride purification) from the production of materials on equipment previously used for the production or manufacturing use (as a reactant, chemical intermediate, or component in a formulating process) of tri- and tetrachlorophenols. (This listing does not include wastes from equipment used only for the production or use of hexachlorophene from highly purified 2,4,5-trichlorophenol).	Tetra-, and pentachlorodibenzo-p-dioxins; tetra- and pentachlorodibenzofurans; tri- and tetra-chlorophenols and their chlorophenoxy derivative acids, esters, ethers, amines, and other salts.
F026	Wastes (except wastewater and spent carbon from hydrogen chloride purification) from the production of materials on equipment previously used for the manufacturing use (as a reactant, chemical intermediate, or component in a formulating process) of tetra-, penta-, or hexachlorobenzene under alkaline conditions.	Tetra-, penta-, and hexachlorodibenzo-p-dioxins; tetra-, penta-, and hexachlorodibenzofurans.
F027	Discarded formulations containing tri-, tetra-, or pentachlorophenol or discarded formulation containing hexachlorophene synthesized from prepurified, 2,4,5-trichlorphenol as the sole component.	Tetra-, penta-, and hexachlorobibenzo-p-dioxins; tetra-, penta-, and hexachlorodibenzofurans; tri-, tetra-, and pentachlorophenols and their chlorophenoxy derivative acids, esters, ethers, amines, and other salts.
F028	Residues resulting from the incineration or thermal treatment of soil contaminated with EPA Hazardous Waste Nos. F020, F021, F022, F023, F026, and F027.	Tetra-, penta-, and hexachlorodibenzo-p-dioxins; tetra-, penta-, and hexachlorodibenzofurans; tri, tetra-, and pentachlorophenols and their chlorophenoxy derivative acids, esters, ethers, amines, and other salts.

In order to treat the wastes banned from land disposal it is necessary to be familiar with both the characteristics and quantities of the wastes which are to be treated. The following sections contain information on the chemical and physical properties of the constituents of concern along with the nature of the waste matrices in which they may be found.

3.2 PHYSICAL AND CHEMICAL CHARACTERISTICS OF CONSTITUENTS OF CONCERN

Chlorinated Dibenzo-p-dioxins

Chlorinated dibenzo-p-dioxins (PCDDs) are organic chemical compounds consisting of two benzene rings connected by two oxygen atoms opposite one another (see Figure 3.1). They may contain one to eight chlorine atoms at any of the eight positions on the two aromatic rings. PCDDs can exist in 75 possible congeneric forms, the most thoroughly studied of which is 2,3,7,8-tetrachlorodibenzo-p-dioxin (TCDD) (U.S. EPA, 1985; Environment Canada, 1985).

Chlorinated dioxins are formed in an exothermic reaction from chlorinated phenols in the presence of base at elevated temperatures. Most chlorophenols and their chlorophenoxy derivatives, whose process wastes are covered under the dioxin listing rule, are manufactured under such conditions. In combustion processes, such as incineration, the mechanism forming TCDD has been estimated to take place most efficiently between 750 and 900°C, while decomposition of TCDD is most likely to occur between 1200 and 1400°C (Ahling, 1977; Junk and Richard, 1981; Redford, 1981; Shaub and Tsang, 1982).

Physical and chemical characteristics of some PCDDs are listed in Table 3-2. Several properties of TCDD are significant with regard to waste treatment. They include the following:

- very low water solubility;

- much greater solubility in organic solvents;

- strong binding to organic matter;

- rapid decomposition at temperatures above 1350°C;

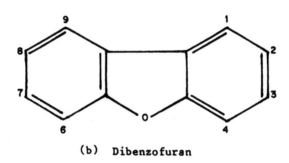

(a) Dibenzo-p-Dioxin

(b)   Dibenzofuran

Figure 3.1.   Structure of (a) Dibenzo-p-Dioxin and (b) Dibenzofuran
Source: Environment Canada, 1985

TABLE 3.2.   PHYSICAL AND CHEMICAL CHARACTERISTICS OF SOME CDDs AND CDBFs

	TCDD	PeCDDs	HxCDDs	TCDF	PeCDFs	HxCDFs
Empirical formula	$C_{12}H_4Cl_4O_2$	$C_{12}H_3Cl_5O_2$	$C_{12}H_2Cl_6O_2$	$C_{12}H_4Cl_4O$	$C_{12}H_3Cl_5O$	$C_{12}H_2Cl_6O$
Molecular weight	322	356	391	306	340	374
Melting point, °C	302-305		240-243	226-228	234-235	247-249
Vapor pressure, estim., 25°C, (mm Hg x 10^6)	0.0001-0.1[f]					
Henry's constant, (atm m^3/mol)	1.5×10^{-6}[f], 2.1×10^{-3}			2.0	1.1	
estim., 25°C, Solubility, ppm		TCDD				
water	1.93×10^{-5}[d], 7.96×10^{-6}[e]	chloroform	370			
benzene	570	acetone	110	benzene 1600[a]		
chlorobenzene	720	methanol	10	toluene 1800		
o-dichlorobenzene	1400	n-octanol	48			
Sorption coefficient (Koc)	1.6×10^6, 4.68×10^5					
Partition coefficients (P):						
hexane/water	10^3					
octanol/water	1.4×10^6					
log P (= log K_{ow})	5.92 - 9.42	9.39 - 10.01	10.22 - 10.89	5.4	6.1	6.7
log K_p (soil/water)	4.59 - 7.09[c]					
Decomposition Temp, °C	700; 1250			1500		
Heat of Combustion, (Kcal/gram)	2.81[b]					

References: USEPA, 1985 (except as noted)

[a] USEPA, 1978.
[b] Federal Register, Volume 51, No. 9.
[c] Jackson, D.R. et al, 1985.
[d] Marple, L. et al., 1986.
[e] Adams, W.J. and D.K. Blaine, 1986.
[f] Crosby, D. G., 1985.

●   low vapor pressure;

●   absorption of ultraviolet radiation; and

●   low rate of biodegradation.

PCDDs are characterized by low polarization which results in a very low
water solubility, but a much higher solubility in organic solvents. The water
solubility of 2,3,7,8-TCDD has been measured by a number of investigators.
Recently derived estimates have been in the range of 7 to 20 parts per
trillion. In contrast, the solubility of TCDD in organic solvents such as
benzene, xylene, and toluene ranges from 500 to 1,800 ppm. This results in a
log octanol/water partition coefficient of up to nine. Consequently, in the
environment TCDD is not generally found at high concentrations in aqueous
media. Instead, it is bound to the organic matter in soil where it may remain
for long periods of time. The half-life of TCDD in soil has been estimated to
range between 1.5 and 10 years (U.S. EPA, 1985), but the results of one recent
study indicated virtually zero degradation of TCDD after being in the soil for
twelve years. In addition, the TCDD had only moved about 10 centimeters over
this period of time (Freeman & Schroy, 1986).
Another property of TCDD is that it absorbs ultraviolet light strongly
with a wavelength of maximum absorption lying within the sunlight region
(above 290 nm). As a result, TCDD has been shown to degrade significantly
when exposed to light of this wavelength in the presence of a hydrogen donor
such as hexane or some other organic solvent. Tests have shown that when a
hydrogen doner is not present, degradation of TCDD is negligible (Crosby,
D. G., 1985). Photolytic degradation was applied to 4,300 gallons of still
bottoms containing 343 ppm of 2,3,7,8-TCDD. The dioxin was first extracted
from the still bottoms using hexane, and then the extract was irradiated with
ultraviolet light; 99.9 percent destruction of TCDD was achieved. This
process will be described in more detail in Section 5.
One of the most important properties of TCDD with respect to treatment is
that it is destroyed at temperatures between 1200 and 1400°C (Shaub and Tsang,
1982). When chlorinated compounds are incinerated at lower temperatures,
however, dioxins may be formed in large quantities. The heat of combustion of
PCDD is 2.81 kilocalories per gram which is greater than the heat of
combustion of compounds such as 1,1,1-trichloroethane and pentachlorophenol.

Consequently, if heats of combustion were used to determine the relative incinerability of hazardous constituents, these compounds could be used as Principle Organic Hazardous Constituents (POHCs) to demonstrate six 9s destruction.

Studies have indicated that the biodegradation of TCDD is a slow process. As mentioned above, the half-life of TCDD in soil has been estimated to range from hundreds of days up to more than ten years (Young, 1976; DiDominico, 1980). One of the reasons that biodegradation is slow is that the organisms cannot readily access the dioxin molecule which is usually strongly adsorbed to soil particles (Crosby, 1985). There has been some recent investigation into the ability of White Rot Fungus to degrade TCDD. This organism is extremely effective in biodegrading lignin. In recent laboratory tests it has also been effective in degrading chlorinated compounds such as dichlorobenzene, DDT and TCDD (Bumpus, J. A., et al., 1985).

Finally, the vapor pressure of TCDD is also very low. A recent estimate of this parameter was 1.5×10^{-7} torr (Freeman, R. A. and J. M. Schroy, 1986). Consequently, volatilization of TCDD from waste streams or soil is not expected to be a rapid process. Of more concern is the wind transport of soil particles with adsorbed TCDD.

Chlorinated Dibenzofurans (CDFs)

Chlorinated dibenzofurans are structurally similar to PCDDs, the only difference being that in PCDFs the two benzene rings are connected by one oxygen atom and one carbon-carbon bond instead of two oxygen atoms. As with PCDDs, the number of chlorine atoms in PCDFs can vary between one and eight, giving rise to 135 possible congeners. Due to the similar structure of PCDDs and PCDFs, the two compounds have similar physical and chemical properties and also show similar toxicity and biological activity (Environment Canada, 1985; U.S. EPA, 1985).

PCDFs are formed as a result of the thermal oxidative cyclization of chlorinated phenols, PCBs, polychlorinated diphenyl ethers, or chlorobenzenes under alkaline conditions. These are similar to the conditions under which PCDDs are formed, and consequently these compounds are frequently found together (U.S. EPA, 1985).

Chlorophenols and Chlorophenoxy Compounds

In general these compounds are water soluble, and in wastes they concentrate in the aqueous phase where they are biodegradable by adapted microorganisms (U.S. EPA, 1985). The biodegradation half-life of 2,4,5-T and Silvex in water is expected to be one to three weeks. The same compounds have a similar half-life due to biodegradation in soils. 2,4,5-TCP, however, has been shown to be persistent in some soils. In one case, where the initial concentration of 2,4,5-TCP in soil was 5000 ppm, the concentration after three years was still 1 to 20 ppm (Lautzenheiser, 1980).

Other properties of these compounds include relatively low volatility, ability to be adsorbed by organic matter such as activated carbon, and susceptibility to photolytic degradation (U.S. EPA, 1985; Lautzenheiser, 1980). Chemical and physical characteristics of these compounds are listed in Table 3-3.

3.3   WASTE SOURCES, QUANTITIES, AND COMPOSITION

3.3.1   Sources of Data

The primary source of data utilized for waste quantity estimates was a report prepared by Technical Resources, Inc. (TRI), entitled, "Analysis of Technical Information to Support RCRA rules for Dioxins – Containing Waste Streams". This report evaluated previous estimates of waste quantities made by Radian (Radian, 1984), and determined that they reflected past practices. TRI presented revised waste generation quantities based on more current information on manufacturing processes obtained by talking to industry contacts. Their estimates appear to be the best available at this time; however, where additional information was available, changes have been made to their estimates.

There are several other sources of data which may be used in the future to obtain better estimates of the quantity of waste containing dioxin. One of these is EPA's Dioxin Strategy. Tiers one, two and three of the Dioxin Strategy encompass sites where 2,4,5-TCP and its pesticidal derivatives were produced or formulated and also sites where wastes from these processes were disposed. Close to 100 potential sites were identified in Tiers 1 and 2

TABLE 3.3. PHYSICAL AND CHEMICAL CHARACTERISTICS OF SOME CHLOROPHENOLS AND CHLOROPHENOXY COMPOUNDS

	2,4,5-TCP	2,4,6-TCP	2,3,4,6-Tetrachlorophenol	Pentachlorophenol	2,4,5-T	2,4,5-TP Silvex	Hexachlorophene
Empirical Formula	$C_6H_3Cl_3O$	$C_6H_3Cl_3O$	$C_6H_2Cl_4O$	C_6HCl_5O	$C_8H_6Cl_3O_3$	$C_9H_7Cl_3O_3$	$C_{13}H_6Cl_6O_2$
Molecular Weight	197.5	197.5	231.9	266.4	255	269.5	406.9
Melting Point, °C	62	68	69	188-191	158	179-181	
Vapor Pressure, mm Hg	400 @ 225°C	60 @ 190°C		1.1×10^{-4} @ 20°C		1×10^{-4}	
Henry's Constant, atm-m^3/mol	7.2×10^{-6}			8.82×10^{-6}			
Vapor Density (relative to air at 25°C)	6.8		8.0	9.20		9.3	
Liquid Density (g/ml)		1.5	1.6	1.98			
Solubility, ppm water acetone methanol	1190	800	insoluble	14	278	140 180 g/kg 134 g/kg	practically insoluble
Sorption Coefficient K_{oc}	19[a]				42[a]	3700[a]	
Partition Coefficients: hexane/water octanol/water (K_{ow}) log K_{ow}	3.06, 3.72	3.38	4.3			2.4	
Decomposition Temp, °C							
Heat of Combustion Kcal/gm	2.88[b]	2.88[b]	2.22[b]		2.88[b]		

References: Handbook of Environmental Data on Organic Chemicals, The Merck Index, The Pesticide Manual.

[a] Lautzenheiser, J. G. et al, 1980.

[b] GCA, 1984.

(Radian, 1984); a report will be issued in the near future containing information related to the extent of contamination at these sites. The information in this report will hopefully contain data which will allow for a better estimate of waste quantities and characteristics, particularly for those sites where waste was disposed (Korb, 1986).

Another potential but unused source of data is RCRA Biennial Reports. These reports are filed biennially by hazardous waste treatment, storage and disposal facilities. EPA has indicated that the data from the 1983 reports are not very accurate both because of a poor response rate, and also because data reported to states were not carefully verified prior to sending data summaries to EPA headquarters (Stoll, 1986). In addition, at the time of the 1983 report, the dioxin waste codes (F020-F028) had not yet been developed. Consequently, the 1983 report only contained data concerning the "U" waste codes for tri-, and tetra-, and pentachlorophenol and their pesticide derivatives 2,4,5-T and Silvex. These waste codes have now been replaced by F027. The 1985 Biennial Reports will contain data on the quantities of waste in codes F020-F028 that were treated, stored or disposed in 1985. These data should be available in the fall of 1986. Whether these data will be better than the data from the 1983 Biennial Reports is unknown at this time.

Finally, facilities that handle (generate, store, treat, or dispose) wastes covered under the dioxin listing rule were required to notify EPA by April 15, 1985. Information contained in the notifications does not at this time include data on the quantities of waste generated or stored, but it does indicate which waste codes the facility handles, and it also includes data to indicate the waste treatment and storage capacity at these facilities. TRI utilized this information to estimate the quantities of F027 waste that will require treatment.

A recent listing of the dioxin waste notifiers is presented in Table 3.4. This listing is updated monthly as new facilities notify, or as facilities that do not belong on the list are deleted. It is expected that information regarding the quantities of wastes handled by these facilities will be assembled in the future.

TABLE 3.4.   LISTING OF DIOXIN NOTIFIERS AS OF FEBRUARY 18, 1986[a]

Waste Code	Company	Location
F020	Monsanto Company	Luling, Louisiana St. Louis, Missouri
	USEPA Laboratory	Kansas City, Kansas
	Velsicol Chemical Corp.	Beaumont, Texas
F021	Koppers Company, Inc.	Montgomery, Alabama Gainesville, Florida Florence, South Carolina Orrville, Ohio Denver, Colorado
	Vulcan Materials Company	Wichita, Kansas
	USEPA Laboratory	Kansas City, Kansas
	Reichhold Chemical, Inc.	Tacoma, Washington
F023	FMC Corporation	Middleport, New York
	Nalco Chemical Corp.	Garyville, Louisiana
	Velsicol Chemical Corp.	Beaumont, Texas
	Monsanto Company	St. Louis, Missouri
	Ralston Purina Health Ind.	Bridgeton, Missouri
	Reichhold Chemical, Inc.	Tacoma, Washington
F026	Monsanto Nitro Plant	Townsend, West Virgina
	Monsanto Company	Luling, Louisiana
	Koppers Company	Montgomery, Alabama Gainesville, Florida Florence, South Carolina Orrville, Ohio Denver, Colorado
	University of Wisconsin	Madison, Wisconsin Arlington, Wisconsin
	USDOE Scientific Laboratory	Los Alamos, New Mexico
	Maytag Company	Newton, Iowa
	Vulcan Materials Company	Wichita, Kansas
	Sunflower Army Ammunition Plant	De Soto, Kansas
F027	Farmland Industries, Inc.	St. Joeseph, Missouri
	Transbag, Inc.	Billings, Montana
	Reichhold Chemical, Inc.	Tacoma, Washington
F028	USDOE Scientific Laboratory	Los Alamos, New Mexico

[a]Does not include commercial treatment facilities which notified because of
their intention to treat these wastes.

3.3.2 Waste Code F020

Sources of Waste--

This waste code includes wastes from the production and manufacturing of
tri- or tetrachlorophenols or intermediates used to produce their
derivatives. The major derivatives include phenoxy compounds such as
2,4,5,-trichlorophenoxyacetic acid (2,4,5,-T), 2-(2,4,5-trichlorophenoxy)
propionic acid (Silvex), and hexachlorophene.

The manufacture of 2,4,5-TCP is accomplished by the alkaline hydrolysis
of tetrachlorobenzene. The primary wastes from the process include
distillation bottoms from solvent recovery, spent filter aids, and reactor
bottoms. These wastes, in addition to the product itself, will be
contaminated with CDDs, CDFs and chlorophenols. The amount of CDDs formed in
the process is dependent upon reaction temperature, which in turn is dependent
upon the solvent used (methanol, ethanol, ethylene glycol, toluene or
isomyl/amyl alcohols). When methanol or water are used as the solvent, the
process operates at around 220-300°C, a temperature at which lab experiments
have shown the formation of 1.6 g TCDD per kg of 2,4,5-TCP. Using ethylene
glycol the process operates at lower temperatures and CDD formation should be
lower (U.S. EPA, 1985).

2,4,6-trichlorophenol and 2,3,4,6-tetrachlorophenol are most efficiently
produced by the chlorination of phenol. In this process more 1,3,6,8-TCDD
than 2,3,7,8-TCDD is formed. These products also contain up to 50 ppm of CDFs.

The manufacture of 2,4,5-T and other phenoxy compounds utilize 2,4,5-TCP
as one of their starting materials. Since TCDD contaminates 2,4,5-TCP, and
may also be generated in the formation of the phenoxy compound itself, it is
expected to be present in both the product and the wastes from its
manufacture. Phenoxy herbicides such as 2,4,5-T and Silvex are synthesized by
reacting the appropriate chlorophenol with a haloalkanoic acid under alkaline
reflux conditions. These conditions are conducive to the formation of CDDs
and CDFs. Careful control of reaction time, temperature,and pH are said to
have an effect in reducing the formation of TCDDs. Wastes from the process
include caustic scrubber water, spent filter aids and/or carbon adsorbent, and
distillation bottoms from solvent recovery. Solvents used are similar to
those used in the production of 2,4,5-TCP. Formerly methanol was used, and
more recently a mixture of ethylene glycol and toluene or xylene was used.

When distillation is used to recover the solvent, CDDs and CDFs are not recovered. They remain either in the still bottoms or the final product. One manufacturer reported the use of carbon adsorption to remove CDDs and CDFs from its product. This, however, results in an additional spent carbon waste.

The manufacture of hexachlorophene is accomplished by condensing prepurified 2,4,5-TCP with formaldehyde in a mixture of concentrated sulfuric acid and ethylene dichloride. Because the reaction occurs at rather low temperature and at acid pH, no CDDs or CDFs are expected to be produced. Past production, when prepurified 2,4,5-TCP was not used, resulted in TCDD contamination due to carry over of contaminants in the TCP feedstock.

Waste Characteristics--

Since these products are no longer manufactured, the wastes requiring disposal are primarily stored in drums or landfills. Most of the wastes, such as those at the Vertac site in Jacksonville, Arkansas were generated as a result of the manufacture of several chlorophenols and chlorophenoxy compounds. At this site, 2,4,5-TCP, 2,4-D and 2,4,5-T were manufactured. Wastes from these processes, primarily toluene still bottoms, were placed in drums which have slowly corroded. These wastes contain toluene (30-50 percent) (Radian, 1984) and lesser amounts of 2,4,5-TCP and tetrachlorobenzene. The concentration of TCDDs in this waste stream, as indicated in Table 3.5, ranges from 0.6 to 350 ppm. The TCDD may not be uniformly distributed throughout the waste liquid. In a similar waste stream, still bottoms from the manufacture of 2,4,5-TCP and hexachlorophene (generated by NEPACCO, stored in Verona, MO, and disposed at Denney Farm), more than 95 percent of the 2,3,7,8-TCDD was incorporated within the solids, which represented only 0.5 percent of the total waste stream (des Rosiers, 1985).

Another type of waste stream covered under waste code F020 is a non-aqueous phase leachate (NAPL) from landfills in which wastes from the production of chlorophenols and phenoxy herbicides have been dumped along with other types of waste including organic solvents. The two major cases of this are the Hyde Park Landfill in Niagra County, New York and the Love Canal Landfill in the City of Niagra Falls, NY. The NAPL from the Hyde Park Landfill contains up to 20.2 ppm of 2,3,7,8-TCDD in addition to "chloro-, bromo-, and fluororganics, PCBs, pesticidal residues, and high concentrations of chromium, antimony, tin, arsenic, lead, zinc, mercury, aluminum,

TABLE 3.5. CONSTITUENTS OF WASTE CODE F020

Source	2,3,7,8-TCDD	TCDDs	Other CDDs	CDFs	Other Possible Constituents	References
Herbicide Manufacture (still bottoms, and reactor residues from the manufacture of 2,4,5-T, 2,4,5-TCP and Hexachlorophene)	60-1290 ppm	0.6-350 ppm			• Methanol • Toluene • Xylene 30-50% • ethylene glycol • ethanol 1-10% • 2,4,5-TCP • trichloroanisoles (methoxyphenyls) 55% • tetrachloro-benzene 0-0.5%	des Rosiers, 1985; Radian, 1984; U.S. EPA, 1985
Non-Aqueous Phase Leachate (from disposal of waste from manufacture of 2,4,5-T and phenoxy herbicides - Love Canal, Hyde Park)	0.2-20.2 ppm				• similar to wastes described above plus: • heavy metals 100-1000ppm – antimony – arsenic – lead – mercury – aluminum – chromium • fluoroorganics • bromoorganics • phosphides • sulfides • PCB	U.S. EPA, 1985
Filter Aids in 2,4,5-TCP manufacture		0.008-300 ppm			• Inorganic Solids 99.5% • chlorophenols 0-0.5%	U.S. EPA, 1985
Filter Cake from Hexachlorophene manufacture		8-2000 ppb			• Inorganic Solids 99.5% • chlorophenols or organic solvents 0.05%	U.S. EPA, 1985
Spent Carbon from treatment of aqueous leachate	untreated waste, 0.004-0.017 ppb; treated waste, 0.010 ppb				• Spent Carbon 99.5% • chlorphenols and other organics 0.05%	U.S. EPA, 1985

phosphorus, and sulfides" (des Rosiers, 1985). Analysis of the NAPL from the
Love Canal Landfill revealed 203 ppb of 2,3,7,8-TCDD. This waste, similar to
the Hyde Park Waste, probably also contains a variety of other constituents
such as heavy metals and PCBs which would influence the selection of an
appropriate treatment method. Control of exhaust gases from the incineration
of such a waste stream would have to take into account the vaporization of
heavy metals present in the waste stream.

Another type of waste covered under this waste code is spent carbon from
the treatment of aqueous phase leachate from Hyde Park and Love Canal. The
aqueous phase of the leachate at Love Canal only contains low levels of
2,3,7,8-TCDD (<1 ppb) but this would probably be concentrated to levels of
above 1 ppb upon adsorption onto the carbon.

Waste Quantities--

The amount of waste currently generated in this code is estimated to be
zero (Technical Resources, Inc. (TRI), 1985). This is based upon the fact
that there has been no manufacture of 2,4,5-TCP since 1983, when Vertac ceased
production. The other major manufacturer of 2,4,5-TCP, Dow, had already
ceased production in 1979. Wastes from the manufacture of 2,4,5-TCP-derived
herbicides such as 2,4,5-T and Silvex are also estimated to be zero since
production of these compounds by major manufacturers was also ceased in 1983.
In addition, future generation of wastes from the manufacture of these
compounds should be zero since EPA issued a draft notice of intent on January
2, 1986 to ban all pesticide products containing 2,4,5-TCP due to their
contamination with 2,3,7,8-TCDD.

Production of hexachlorophene at this time is also zero due to the
unavailability of 2,4,5-TCP necessary for its manufacture. It is possible,
however, that hexachlorophene will be manufactured in the future using a new
dioxins and furans free process that Velsicol Chemical Corp. is interested in
developing.

Since there are no current manufacturing sources of waste in this code,
the major sources of wastes requiring treatment are wastes which were
generated in the past and are currently being stored. The quantity of waste
designated as F020 is summarized in Table 1.2. It includes 2273 metric
tons (MT) of still bottoms at the Vertac site in Arkansas, 1364 MT of NAPL at
the Hyde Park Landfill, 68 MT of NAPL at the Love Canal Landfill, and 23 MT of

spent carbon from the treatment of aqueous phase leachate from the Love Canal
Landfill.   The TRI report did not include an estimate of the quantity of spent
carbon from treatment of aqueous phase leachate at the Hyde Park Landfill.
EPA Region II personnel have indicated that there are currently several
dumpsters of spent carbon at the site that Calgon (the manufacturer of the
carbon) will not accept for regeneration because of possible dioxin
contamination.   Aqueous phase leachate is still being generated at a rate of
approximately 5000 gallons per day, so spent carbon will also continue to be
generated and require treatment (Gianti, 1986).

In addition to the non-aqueous phase leachate that is currently stored in
lagoons at Hyde Park, an additional 40 to 200 gallons per day are continuing
to be collected.   At an average rate of 120 gallons per day, 200 MT would be
collected in one year.   The period of time over which NAPL will continue to be
generated is not known (Gianti, 1986).

In addition to these major sources of waste, there are probably other
smaller sites where 2,4,5-TCP and derivatives were manufactured or formulated
in the past and wastes are currently stored.   Possible locations of these
sites have been and are being identified through EPA's Dioxin Strategy and
also through the provision that all potential handlers of dioxin wastes notify
EPA of their activities.   Data on the quantity of wastes at these "other"
sites, however, are not currently available.   TRI estimated that the quantity
of FO20 waste at these miscellaneous sites is 500 MT.

3.3.3   Waste Code FO21

Sources of Waste--
This waste code encompasses wastes from the production and manufacturing
of pentachlorophenol.   Pentachlorophenol (PCP) and its sodium salt have
various uses as fungicides and biocides with the majority (80 percent) being
used as a wood preservative.   Since all non-wood uses of PCP will be banned as
a result of a notice of intent to cancel made by EPA on January 8, 1986
(Chemical Regulation Reporter, 1/10/86), it is expected that all future uses
will be as a wood preservative.

The manufacture of pentachlorophenol can be accomplished either by the
chlorination of phenol or by the alkaline hydrolysis of hexachlorobenzene.   In
the United States, the former method is used.   The chlorination usually

proceeds until approximately 3 to 7 percent tetrachlorophenol remains in the product. The chlorination reaction results in the formation of up to 30 ppm of HCDDs along with chlorinated furans. 2,3,7,8-TCDD, is not expected to be generated in the manufacture of PCP (U.S. EPA, 1985).

Current manufacture of PCP does not generate any dioxin-containing wastes (TRI, 1985). The primary reason for this is that distillation is not part of the process. At one time, however, Dow did manufacture a purified PCP product. Purification of the product was achieved by distillation, and the still bottoms that were generated contained up to 2,000 ppm of CDDs and CDFs. Because of difficulties in working with the product, however, its production was soon dropped (TRI, 1985; U.S. EPA, 1985).

The future manufacture of pentachlorophenol, however, may involve purification and an associated waste stream containing CDDs and CDFs. A regulation under the Federal Insecticide, Fungicide and Rodenticide Act (FIFRA) would require PCP manufacturers to reduce the concentration of HxCDD in PCP from a current average of 15 ppm to 1 ppm (51 FR 1434). PCP manufacturers are disputing this regulation, and negotiations between them and EPA are currently being carried out to resolve the HSVE. Once this requirement takes effect, there will be an ongoing dioxin waste stream associated with the manufacture of PCP unless new manufacturing processes are developed in which the formation of dioxins is completely avoided (Chemical Regulation Reporter, 1/10/86).

Another possible source of waste in this code is from the formulation of pentachlorophenol products. Eighty percent of all PCP, however, is used for wood preservation, where no formulation of PCP is required. Non-wood uses of PCP have been banned. Consequently, wastes from the formulation of PCP are not expected to be generated in the future.

The final potential source of FO21 waste is the waste from wood treatment/preservation facilities that use pentachlorophenol. This waste is not currently covered by the dioxin listing rule. It has been surmised, however, that this may be the largest source of PCDD/PCDF contaminated waste (des Rosiers, 1985). This waste is currently designated as RCRA code K001-Wastewater Treatment Sludge from Wood Preserving Processes using PCP. The wastewater treatment sludge would most likely be contained in a lagoon in which other organic wastes in addition to PCP may have been placed. In many cases these lagoons were torched to reduce waste volumes, resulting in the

formation of PCDDs and PCDFs (des Rosiers, 1985). The number of sites
containing these sludges may exceed 100; however, no estimates of the quantity
of sludges has been made. In addition, 23 damage incidents related to wood
treating operations using PCP were included in the Listing Background
Document. These damage incidents include cases where sludges and wastewaters
were stored onsite and contaminated soil and water. This waste stream is
being included in this discussion due to its possible listing as a "dioxin
waste" to be banned from land disposal.

Waste Characteristics--

As mentioned above, the manufacture of PCP does not currently generate a
waste stream containing CDDs. In the future, however, PCP will have to be
purified to reduce HCDD concentrations from 15 ppm to 1 ppm. Radian estimated
the composition of a waste stream generated as a result of the purification of
PCP by distillation (Radian, 1984). This waste stream would consist primarily
of organic solids (nonvolatile), various chlorinated phenols and organic
solvents as indicated in Table 3.6. In addition, a small fraction of the
waste would consist of residual catalyst (aluminum chloride), and total CDDs
could reach 2,000 ppm. Future purification of PCP will probably not be by
distillation, but instead by a solvent extraction and crystallization
process. The wastes from this process are assumed to be similar to those
generated by distillation (TRI, 1985).

Waste from wood treatment facilities is expected to contain a large
variety of constituents. The exact composition of the waste will vary from
facility to facility, but in all cases will be a sludge with varying
concentrations of water, chlorophenols, and creosote. Organometallic
compounds such as copper and zinc naphthalenates and arsenicals are also
expected to be potential constituents, in addition to PCBs and waste
solvents. The source of CDDs and CDFs in these wastes would be from the
inherent contamination of PCP with these compounds. Higher concentrations may
be present if waste pits and lagoons containing these wastes were torched to
reduce volume.

TABLE 3.6.   CONSTITUENTS OF WASTE CODE F021

Source	CDDs	CDFs	Other Potential Constituents		Reference
Purification of Technical Grade PCP (By Distillation[a])	up to 2000 ppm	should be present but not quantified	• non-volatile organic solids	60-90%	Radian, 1984;
			• chlorinated or aromatic solvents	15-30%	U.S. EPA, 1985
			• chlorinated phenols	0-5%	
			• aluminum trichloride	0-2%	
			• aluminum hydroxide	0-2%	
Waste from Wood Preserving (K001)	b	b	• water	60-98%	IT Enviroscience, 1981;
			• 2-chlorphenol	up to 8000 ppm	ICF, Inc., 1984;
			• pentachlorphenol	up to 6000 ppm	des Rosiers, 1985
			• phenol		
			• creosote		
			• benzo(a)pyrene		
			• 2,4-dimethylphenol		
			• fluoranthane		
			• arsenicals		
			• copper and zinc napthenalates		
			• PCBs		

[a]Waste from future purification is assumed to be similar to past purification by distillation.

[b]Data on content of CDDs and CDFs are not currently available.

Waste Quantities--

The only current manufacturer of PCP is Vulcan Materials Company.
Previous manufactures include Dow, which ceased production in 1980, and
Reichhold Chemicals, Inc. which ceased production in 1985 (TRI, 1985). TRI
estimated that Vulcan would fill the current U.S. demand for PCP of 15,000 MT
per year. If this is the case, and if purification results in a waste stream
of 5 percent of the end product, the quantity of waste generated through
purification will be 750 MT per year.

Wastes from formulation of PCP are not expected to be generated in the
future since all PCP is expected to be sold directly to wood preservers.
Previous to the ban of non-wood uses of PCP, however, 20 percent of the PCP
was formulated into products for herbicidal, antimicrobial, and disinfectant
use (Chemical Regulation Reporter, 1/10/86). These uses are assumed to have
resulted in the generation of 350 MT per year of scrubber water sludges
contaminated with PCP and HCDD. Three hundred fifty MT are estimated for the
past years (1985) formulation activities, and another 350 MT for the current
years activities. Therefore, as indicated previously in Table 1.2, 700 MT of
waste code F021 presently require treatment.

3.3.4 Waste Code F022

The Radian Report (Radian, 1984) states that there are no known
commercial activities with the processes encompassed by this waste code. The
compound, 2,4,5-trichlorophenol, was manufactured by the alkaline hydrolysis
of tetrachlorobenzene which would subject its wastes to inclusion in this
waste code, in addition to waste code F020. Since 2,4,5-TCP is not being
manufactured at this time, no wastes are currently being generated.
Potentially generated wastes, and wastes generated by this process in the past
are discussed previously under waste code F020.

3.3.5 Waste Code F023

Sources of Waste--

Production trains are often used for the production of chemicals whose
manufacture necessitates the use of similar process equipment. In the
manufacture of chemicals on a production train previously contaminated with

PCDDs, both the products and the wastes may be contaminated with PCDDs. Such is the case with the manufacture of 2,4-D, which from process chemistry is not otherwise expected to be contaminated with 2,3,7,8-TCDD, but does contain TCDD due to the use of equipment previously used to produce 2,4,5-T (Federal Register, 1980). The continued manufacture of 2,4,D on 2,4,5-T contaminated equipment would theoretically extract the residual 2,4,5-T and TCDD over time. There are data, however, which shows that still bottoms generated seven years after 2,4,5-T production ceased, still contained 70 ppb of TCDD (U.S. EPA, 1985).

Waste Characteristics--

The types of waste generated under this waste code would be similar to those generated under waste code F020, and would include still bottoms, reactor residues, and filter aids. In addition, TRI stated that most of the waste under this code would likely come from formulating processes using equipment previously used to formulate chlorophenols and their phenoxy derivatives. If this is the case, washwater sludges and other equipment cleanup wastes would also be potential waste types.

Sampling and analysis of 2,4-Dichlorophenol still bottoms from a 2,4-D manufacturing facility where 2,4,5-TCP had been manufactured years earlier showed 20 ppm of CDDs and 450 ppm of CDFs (U.S. EPA, 1985).

Waste Quantities--

TRI estimated the quantity of waste generated under this code to be zero. This estimate is based on several factors. One is the assumption that the number of formulators of products covered by the code has been reduced substantially, and much of the equipment of interest has been replaced, dismantled, sold or discarded. Another is that most of the facilities that notified EPA that they were generating F023 also indicated that they treated their wastes onsite, and therefore would not present a demand on offsite waste treatment facilities. Finally, they indicated that, for facilities that also generate other listed dioxin wastes, it would be difficult to differentiate these wastes from the F023 waste.

Radian also made an estimate of the quantity of waste generated under this code. Their estimate is based on the assumption that past manufacturers and formulators of the products of concern, characterized as active, are currently using contaminated equipment and generating contaminated wastes. In addition, they assumed that current production levels are equal to the greatest past production levels. These assumptions would seemingly lead to an upper bound estimate of the quantity of F023 waste generated. The actual quantity probably lies between TRI's estimate of zero and Radians estimate of 573 MT per year.

3.3.6 Waste Code F026

The only manufacturing process that involves the manufacturing use of tetra-, penta-, or hexachlorobenzenes is the manufacture of 2,4,5-Trichlorophenol. As mentioned above, the manufacture of 2,4,5-TCP involves the alkaline hydrolysis of tetrachlorobenzene. Wastes from the production of materials on equipment previously used to manufacture 2,4,5-TCP, however, is regulated under waste code F023. Consequently there should be no F026 waste generated..

3.3.7 Waste Code F027

Sources of Waste--
This waste code encompasses discarded, unused formulations of tri-, tetra-, and pentachlorophenols and their derivatives. These wastes arise either because the product is off specification, the product was manufactured but then its use was banned, or an excess amount was produced or acquired. Because most of these compounds are no longer being manufactured, these wastes are not currently being generated. The exception to this, as mentioned above, is pentachlorophenol which is still manufactured, and so wastes from unused formulations may continue to be generated. For the other compounds of concern unused formulations which have been generated in past years may still lie in storage and eventually require final destruction/disposal.

Waste Characteristics--
Measured concentrations of CDDs and CDFs in the products of concern are presented in Table 3.7. As indicated by the data in this table, the concentration of CDDs in these products can range from non-detectable levels

TABLE 3.7.   PCDD AND PCDF CONCENTRATIONS IN MANUFACTURED PRODUCTS AND CHEMICAL INTERMEDIATES (ppb)

Source	2,3,7,8-TCDD	TCDDs	PeCDDs	HCDDs	PCDDs	PCDFs	Physical Form of Formulations
Trichlorophenol	1.8-6200	<.02-49,000	ND-1500	ND-10,000	<30	60,000	as a sodium salt
Tetrachlorophenol		ND-<700	ND-5200	41,000-100,000	<22,000-<100,000	160,000-865,000	
Pentachlorophenol: o techincal/commercial		ND-120	ND-30	ND-38,500	ND-520,000[a]	80-780,000	as itself (crystals); solutions in oil; as a water soluble sodium salt; mixed
o purified (Dow's EC-7)[d]	50			1000		3400 (HCDF)	with creosote
2,4,5-T	100-90,000[b]	10-10,000		>10,000			as an emulsifiable concentrate 4-6 lbs/gallon active ingredient;
Silvex		100-10,000		500			emulsifiable concentrate,
Hexachlorophene	30	100		10-390[c]			emulsifiable concentrate, 25% AI
Agent Orange	100-47,000						1:1 mixture of butyl esters or 2,4,5-T and 2,4-D
Agent Purple	32,800[e]						50% n-butyl 2,4-D, 30% n-butyl 2,4,5-T, 20% isobutyl 2,4,5-T

Reference: U.S. EPA, 1985; Rappe, 1979; Young, 1983.

[a]Produced in 1975.

[b]Used for Agents Orange and Purple.

[c]Produced in 1978.

[d]88% Penta, 12% Tetra.

[e]Produced in the '60s.

up to hundreds of ppm. The actual concentration will vary from batch to batch, and compound to compound. In addition, the products may have contained higher concentrations of CDDs in the past than they have more recently. For example, the mean concentrations of 2,3,7,8-TCDD in Agent Orange and Agent Purple (both mixtures of 2,4,5-T and 2,4-D) in the 1960s were 1.98 and 32.8 ppm respectively (Young, 1983) while it was claimed that those prepared in the 1970s contained less than 0.1 ppm (Rappe, 1979). Despite this claim there may still remain quantities of waste in storage with substantial levels of TCDD. For example, the Tennessee Valley Authority (TVA) currently has 21 drums of herbicide orange, one of which contains 5.6 ppm of 2,3,7,8-TCDD (TRI, 1985).

The only F027 waste being generated on an ongoing basis would be unused pentachlorophenol. Because of the consent decree requiring PCP manufacturers to reduce the concentration of HCDD in their product from 15 ppm to 1 ppm, this waste will be of less concern than it has been in the past, since the majority of the HCDD will be incorporated in the purification wastes.

The physical forms of these wastes will vary from case to case. Pentachlorophenol is commonly applied to wood as a 5 percent suspension in fuel oil (sometimes blended with creosote) or dissolved in an organic solvent (Chemical Products Synopsis, 1983). The other products are generally marketed as emulsifiable concentrates. These concentrates are prepared by dissolving the active ingredient (15-80 percent) and a surface active agent (less than 5 percent) in a water emulsifiable organic solvent. The surface active emulsifiers are generally polyethylene and polypropylene glycols, calcium sulfonates or various soaps (Sitig, 1980).

Waste Quantities--

TRI estimated the quantity of waste code F027 in storage to be 1,000 MT per year. This estimate was based on a review of data contained in the Dioxin Waste Notifications. They estimated that the amount of waste generated that would place a demand on treatment capacity was equivalent to the storage capacity at facilities notifying EPA that they were handling waste code F027. Their estimate does not include facilities which reported capacity for waste treatment within the plant site because the waste generated at these facilities would not place a demand on offsite treatment.

Current and future generation of this waste code should only be associated with the manufacture of PCP (unless the manufacture of 2,4,5-TCP resumes). One thousand MT appears to be the maximum annual amount that would be generated. It has been reported, however, that the quantity may be much less (Industrial Economics, Inc., 1986). Since Reichhold Chemicals recently stopped manufacturing PCP, the demand is greater than the capacity for production. Consequently, the quantity of unused PCP may be less that it has been in the past.

3.3.8 Contaminated Soils

Waste Sources--

Soils contaminated with listed dioxin wastes are regulated as the respective hazardous waste contaminant (50 FR 28713). These wastes are categorized in one of the waste codes discussed above. They are being discussed in this section, however, because of the difference in the physical nature of these wastes relative to the other listed wastes, and also because it is difficult to assess exactly which waste code is appropriate for the contaminated soil.

The largest known quantities of contaminated soils are in Missouri where several horse arenas and other areas were sprayed with TCDD contaminated waste oils. Contamination of the waste oils with TCDD resulted from mixing these waste oils with distillation bottoms from the manufacture of 2,4,5-trichlorophenol to be used to produce hexachlorophene (USEPA, 1985).

Other known sources of contaminated soil and sediment include a herbicide manufacturing plant (Vertac Chemical Company) in Arkansas where cooling pond sludges, equalization basin muds, and stream sediments have been contaminated due to the leakage of wastes (such as still bottoms from the manufacture of 2,4,5-TCP) from drums stored onsite (Thibodeaux, L.J., 1983). The improper disposal, both onsite and offsite, of wastes from another herbicide manufacturer in Missouri have also led to the contamination of soils in that state. Instances of improper disposal of waste from this site include dumping of drummed and bulk wastes into unlined onsite trenches and lagoons, and the disposal of drummed wastes into a trench on a farm offsite (U.S. EPA, 1985).

Large quantities of contaminated soils and sediments also exist in New York State. Two landfills, Hyde Park and Love Canal, were used for the disposal of organic solvents and wastes from the production of chlorophenols and phenoxy herbicides. The Hyde Park landfill is estimated to contain approximately 120 kg of TCDD; leakage of wastes from these landfills has resulted in the contamination of surrounding soils and sediments. It is believed that there are 55,000 cu yds of stream bed sediments contaminated with an average of 70 ppb of TCDD (USEPA, 1985).

Waste Characteristics--

Waste soils, sediments and other solid materials that have been contaminated with dioxin may have varied compositions and concentrations of dioxin. As indicated in Table 3.8 concentrations of TCDD in soils and other solids range from nondetectable (ND) to greater than 26 ppm. The contaminated materials at some of the sites include not only granular materials such as soils and sand, but also asphalt, vegetation, rocks and gravel. These materials may require special procedures to remove and/or destroy the TCDD. Incinerators many times require the feed material to be of a certain size and consistency. Consequently, some sort of pretreatment to reduce size and produce a more uniform feed may be necessary for the treatment of these wastes.

One of the most significant characteristics of TCDD on soils is its very high soil/water partition coefficient. As shown in Table 3.2 the log of the partition coefficient can be as high as seven. The conclusions of a recent study indicate that the most important factor affecting both the concentration of TCDD in soils, and its partitioning between soil and water is the presence of other organics in the soil. The data indicated that in soils with higher concentrations of solvent-extractable organics (particularly halogenated semivolatiles) the TCDD concentrations in water extracts were greater. They further suggested that it is other organic contaminants in the wastes and not the total organic carbon and clay content of soils that affects the mobility of TCDD (Jackson, D. R. et al, 1985). This would mean that in cases where wastes from the manufacture of chlorophenols or chlorophenoxy herbicides were disposed or leaked into soil media along with other organic wastes, TCDD may be much more mobile than would be normally expected. In cases where chlorophenols and organic solvents are not present in the wastes, the TCDD may be much more strongly bound to the soil, and much more difficult to desorb.

TABLE 3.8.   CHARACTERIZATION OF SOME SOILS CONTAMINATED WITH DIOXINS CONSTITUENTS (ppb)

Location	2,3,7,8-TCDD	TCDDs	Waste Matrix
Herbicide Manufacturing Plant, AR[a] (Vertac)		ND-2.9 ND-1.8 22.1 2.1 0.55-53	Soil Creek Sediment Pond Sediment Stream and Drainage Ditch Sediments
Hyde Park[b]	0.7-4.5	0.4-220	Stream Sediments
Love Canal[b]	ND-9570 ND-655		Sump Sediment Stream Sediment
Missouri Soils[c]	8-2438		From 8 Missouri Sites; soils were silty loams and sandy loams ranging from 38.5 to 72.1% sand, 30.5 to 51.6% silt, and 2.3 to 12.5% clay. Organic matter ranged from 15 mg/g to 76 mg/g
New Jersey Soils[c]	1083; 26,316		One soil was 72% sand, 23% silt, 5% clay, 37 mg/g organic; the other was 81% sand, 18% silt, 1% clay, 80 mg/g organic

[a]Thibodeaux, L.J., 1985.

[b]U.S. EPA, 1985.

[c]Jackson, D.R. et al, 1985.

Waste Quantities--

It has been estimated that there are 500,000 MT of dioxin-contaminated soils in Missouri, 160,000 MT at Times Beach alone (Radian, 1984). Radian made a rough estimate of the total quantity of dioxin-contaminated soil in the U.S. by assuming an average site size (5 acres with a 1.5 ft. depth) and multiplying this by the number of tier 1, 2, and 3 sites identified in the Dioxin Strategy. Their result was 2.3 million MT. Because of the uncertainties in making an estimate such as this, 500,000 MT was used as the minimum quantity of dioxin-contaminated soils currently requiring treatment in the United States. This is the number that is presented in Table 1.2. It will be possible to make a more accurate estimate of the quantity of contaminated soil after the sites identified in the Dioxin Strategy are better characterized.

Wastes which are of concern for this document are those containing an extractable TCDD or TCDF concentration of greater than 1 ppb. It is possible that much of the contaminated soil will contain strongly adsorbed TCDD, and so will not require treatment with respect to the land disposal ban.

REFERENCES

Adams, W.J. and D.K. Blaine. A Water Solubility Determination of
2,3,7,8-TCDD. Paper presented at the fifth International Symposium on
Chlorinated Dioxins and Related Compounds, Bayreuth, Germany,
September 16-19, 1985.

Ahling, B. et. al. 1977. Formation of polychlorinated dibenzo-p-dioxins
and dibenzofurans during combustion of a 2,4,5-T formulation.
Chemosphere 8:461-468.

Bumpus, J.A. et. al. Biodegradation of Environmental Pollutants by The
White Rot Fungus Phanerochaete Chrysosporium. Presented in the Eleventh
Annual Research Symposium on Incineration and Treatment of Hazardous
Waste. EPA 600/9-85-028

Chemical Regulation Reporter, January 10, 1986.

Crosby, D.G. The Degradation and Disposal of Chlorinated Dioxins
presented in proceedings of symposium entitled: Dioxins in the
Environment. Edited by Kamrin, M.A. and P.W. Rodgers, 1985.

des Rosiers, P.E. Memorandum to Erich Bretthauer, Chairman, ORD Dioxin
Team. June 18, 1985.

DiDominico, A. et. al. Accidental Release of 2,3,7,8-TCDD at Seveso,
Italy. Ecotoxicology and Environmental Safety, 4 (3) 282-356, 1980.

Environment Canada. Polychlorinated Dibenzo-p-Dioxins (PCDDs) and
Polychlorinated Dibenzo-Furans (PCDFs): Sources and Releases. Prepared
by A. Szheffield, Environmental Protection Service. July, 1985. EPS
5/HA/Z

Federal Register, 1980. Storage and Disposal of Waste Material:
Prohibition of Disposal of Tetrachlorodibenzo-p-dioxin. 45 (98): 32676.

Freeman, R.A. and J.M. Schroy. Modeling of the Transport of 2,3,7,8-TCDD and Other Low Volatility Chemicals in Soils Environmental Progress, 5 (1), 1986.

GCA Technology Division. Assessment of Treatment Alternatives for Wastes Containing Helogenated Organics. U.S. EPA Contract No. 68-01-6871. October, 1984.

Gianti, S. U.S. EPA, Region II. Telecon with M. Arienti, GCA Technology Division. March 6, 1986.

ICF, Incorporated. The RCRA Risk/Cost Analysis Model Phase III Report. Submitted to U.S. EPA Office of Solid Waste, Economic Analysis Branch. March 1, 1984.

Industrial Economics, Inc. Regulatory Analysis of Proposed Restrictions on Land Disposal of Certain Dioxin-Containing Wastes. Draft Final report prepared for EPA, Office of Solid Waste. January, 1986.

IT Enviroscience, Inc. Study of Potentially Hazardous Waste Streams for the Industrial Organic Chemical Manufacturing Industry, 1982.

Jackson, D.R. et. al. Leaching Potential of 2,3,7,8-TCDD in Contaminated Soils. In proceedings of the Eleventh Annual Research Symposium on Land Disposal of Hazardous Waste. EPA/600/9-85/013 April 1985.

Junk, G.A. and J.J. Richard. 1981. Dioxins Not Detected in Effluents from Coal/Refuse Combustion. Chemosphere 10:1237-1241.

Korb, Barry. U.S. EPA. Telecon with M. Arienti, GCA Technology Division. February 10, 1986.

Lautzenheiser, J.G. et. al. Non Aquatic Fate and Environmental Bruden of 2,4,5-T, 2,4,5-TP and 2,4,5-TCP. Prepared for U. S. EPA Contract No. 68-01-3867. September, 1980.

Marple, L. et. al. Water Solubility of 2,3,7,8-Tetrachlorodibenzo-p-dioxin. Environmental Science and Technology. 20 (2) 1986.

Mill, T. SRI International, Menlo Park, CA

Radian Corporation. Assessment of Treatment Practices For Proposed Hazardous Waste - Listings FO20, FO21, FO22, FO23, FO26, FO27, and FO28. EPA Contract No. 68-02-3148. September, 1984.

Rappe, C. et. al. Dioxins, Dibenzofurans, and other Polyhalogenated Aromatics Production, Use, Formation and Destruction. Ann. NY Acad. Sci. 320: 1-18.

Redford, D.P. et. al. 1981. Emission of PCDD from combustion sources. International Symp. on Chlorinated Dioxins and Related Compounds. Arlington, VA; October 25-29.

Young, A.L. et. al. Fate of 2,3,7,8-TCDD in the Environment: Summary on Decontamination Reccomendations. U.S.A. FA-TR-76-18, 1976.

Sittig, M. Pesticide Manufacturing and Toxic Materials Control Encyclopedia. Noyes Data Corporation, Park Ridge, NJ, 1980.

Shaub, W.M. and W. Tsang. 1982. Physical and Chemical Properties of Dioxins in Relation to Their Disposal. Proc. Second Intl. Symp. on Dioxins. Arlington, VA, October. 1981.

Stoll, Barry. U.S. EPA Office of Solid Waste. Telecon with M. Arienti, GCA Corporation Technology Division. February 25, 1986.

Technical Resources, Inc. Analysis of Technical Information to Support RCRA Rules for Dioxins-Containing Waste Streams. Submitted to P.E. des Rosiers , EPA/Office of Research and Development. Contract No. 5W-6242-NASX. July 31, 1985.

Thibodeaux, L.J. Offsite Transport of 2,3,7,8-Tetrchlorobenzo-p-dioxin from a Production Disposal Facility. In: Chlorinated Dioxins and Dibenzofurans in the Total Environment. C. Gangadhar et. al. Ed. Butterworth Publishers. pp. 75-86, 1983.

U.S. EPA, 1978. Report of the Ad Hoc Study Group on Pentachlorophenol and Contaminants. EPA/SAB/78/001 p.170

U.S. EPA Dioxin Listing Background Document. U.S. EPA, OSW, Washington, D.C., January, 1985.

Westat, Inc. Natural Survey of Hazardous Waste Generators and Treatment, Storage and Disposal Facilities Regulated Under RCRA in 1981. Prepared for EPA Office of Solid Waste, 1983.

Young, A.L. Long-Term Studies on the Persistence and Movement of TCDD in a Natural Ecosystem. In: Human and Environmental Risks of Chlorinated Dioxins and Related Compounds. R.E. Tucker et. al., Ed. Plenum Publishing Corp., NY pp. 173-190.

4. Thermal Technologies for Listed Dioxin Wastes

In this report, thermal technologies include incineration, pyrolysis and other processes in which heat is the major agent of destruction. As mentioned in Section 3, laboratory studies have shown that CDDs break down rapidly when subjected to temperatures above 1,200°C. As a result, high temperature incineration and other thermal methods have received much attention with regard to treatment of waste containing CDDs. This attention has led to the development by EPA of a mobile incineration system designed specifically for research on wastes containing dioxin and other toxic substances. This mobile incinerator has demonstrated greater than six nines (99.9999 percent) destruction and removal efficiency (DRE) of wastes containing CDDs, and has led EPA to propose in their January 14, 1986 ruling on land disposal of waste containing dioxins (FR, Vol. 51, No. 9) that incineration (or an equivalent thermal technology) be used as the treatment technology for these wastes.

Incineration and other thermal treatment of RCRA-listed dioxin wastes (codes FO20, FO21, FO22, FO23, FO26, FO27) must be done in accordance with the criteria specified under 40 CFR Parts 264.343 and 265.362 in the dioxin listing rule. These criteria specify that processes burning these wastes must achieve a DRE of 99.9999 percent for each principal organic hazardous constituent (POHC) designated in its permit. DRE is determined from the following equation:

$$DRE = \frac{(W_{in} - W_{out})}{W_{in}} \times 100$$

where: W_{in} = mass feed rate of one POHC in the waste stream feeding the incinerator; and

W_{out} = mass emission rate of the same POHC present in exhaust emissions prior to release to the atmosphere.

Six nines (99.9999 percent DRE) must be demonstrated either on the CDDs themselves or on a POHC that is more difficult to destroy than the CDDs. One criterion EPA has used to determine the relative ease of destruction of various toxic constituents is heat of combustion. The lower the heat of combustion, the more difficult it is to destroy. Therefore, if a waste containing HCDD is to be destroyed, the process must be able to achieve six nines DRE on a POHC with a heat of combustion of less than 2.81 kcal/gram, the heat of combustion of HCDD. An example of such a compound would be pentachlorophenol, which has a heat of combustion of 2.09 kcal/gram.

Another criterion for evaluating incineration (or any treatment process) with regard to dioxin wastes is that the residues of treatment must contain less than a detectable level (1 ppb) of CDDs and CDFs in order to be designated non-hazardous. Consequently, when evaluating the potential of some method for treating wastes containing dioxin, it is important not only to look at the exhaust gases but also the scrubber water, filter residues, and the non-combusted, treated material if the waste is an inorganic solid such as soil.

Finally, with regard to destruction by incineration, CDDs and CDFs are assumed to be similar to polychlorinated biphenyls (PCBs). This is because they are both highly chlorinated compounds with similar structure and similar heats of combustion. In addition, incinerators burning PCBs must achieve six nines DRE. As a result, EPA has indicated that incinerators that operate in accordance with the performance standards specified in 40 CFR 761.70 for PCB wastes, namely six nines DRE, have also demonstrated their ability to meet the performance standards for incinerators burning dioxin wastes (51 FR 1602).

Incinerators burning PCB wastes must be operated at 1200°C with a waste residence time of 2 seconds and 2 percent excess oxygen. Alternatively, the incinerator may be operated at 1600°C, 1.5 second dwell time and 2 percent excess oxygen. It would be expected that incinerators burning dioxin wastes would also have to operate under these conditions.

This section includes subsections on a variety of thermal technologies. Methods of incineration include:

- Stationary Rotary Kiln
- Mobile Rotary Kiln
- Liquid Injection
- Fluidized Bed
- Infrared

Other thermal destruction technologies include:

- High Temperature Fluid Wall
- Plasma Arc
- Molten Salt
- In Situ Vitrification
- Supercritical Water Oxidation

Each subsection contains a process description, an evaluation of the performance of the technology with regard to chlorinated dibenzo-p-dioxins (CDDs) or similar compounds, an assessment of treatment costs, and a discussion of the status of the technology. Not all of these units have been tested using dioxin waste, but most of them have at least been tested using PCB waste; in these cases, the PCB data have been presented as evidence of their performance.

4.1   STATIONARY ROTARY KILN INCINERATION

Several commercial rotary kilns have been permitted to burn PCB wastes. In so doing they have demonstrated six nines DRE for PCBs, and therefore have the potential to burn dioxin wastes. These units are: the Rollins incinerator in Deer Park, Texas; the SCA incinerator in Chicago, Illinois; and the ENSCO incinerator in El Dorado, Arkansas. None of these units has been demonstrated using dioxin wastes; however, the EPA Combustion Research Facility in Jefferson, Arkansas, which operates a rotary kiln incinerator, recently conducted test burns of dioxin wastes. Even though this is not a commercial incineration facility, the data that were generated by the dioxin burns are included to indicate the performance of a rotary kiln.

4.1.1   Process Description(GCA, 1985; McGaughey, et al. 1984; Bonner,
        1981; Freeman and Olexsey, 1986)

A rotary kiln incinerator consists of a cylindrical, refractory-lined
shell that is mounted with its axis at a slight incline (less than 5 degrees
from the horizontal) and may rotate from 5 to 25 times per hour. The
peripheral speed of rotation, which ranges from 1 to 5 feet per minute,
provides excellent mixing of wastes and combustion air. In addition to the
kiln, the system includes a waste feed system, a secondary combustion chamber,
air pollution control equipment, and a stack. Rotary kilns can handle a wide
variety of waste feeds. Solids and viscous sludges are typically fed to the
upper end of the kiln by conveyor or in fiber drums, while liquids are
atomized through auxiliary burners. Liquids are either injected into the kiln
or into the afterburner. In the kiln, solid wastes are partially burned. The
products are gases and inorganic ash. The ash is removed from the kiln, and
combustion of the gaseous products is completed in the secondary combustion
chamber (afterburner) (Bonner, T. A., et al., 1981; Marson, L. and S. Urger,
1979; McInnes, R. B., 1979.)

Auxiliary fuel systems are typically required to bring the kiln up to the
desired operating temperature. Various types of auxiliary fuel systems may be
used, including dual-liquid burners designed for combined waste-fuel firing or
single-liquid burners equipped with a premix system.

Rotary kilns may be configured either with a co-current or a
countercurrent design. Co-current units have the auxiliary fuel burner at the
same end of the incinerator as the waste feed, whereas countercurrent units
are designed such that the combustion gases run countercurrent to the flow of
waste through the incinerator. The countercurrent design is more advantageous
for wastes having a low heating value because temperature can be controlled at
both ends of the kiln, minimizing problems such as overheating of the
refractory lining. Brief descriptions of two of the rotary kiln incineration
facilities that are permitted to burn PCBs are provided below.

Rollins (Rollins, 1985; M. M. Dillon, 1983; Gregory, 1981)--

    The configuration of the Rollins stationary incinerator is shown in
Figure 4.1.1. Solids or sludges are conveyed to the rotary kiln in fiber
drums or 55 gallon metal drums. Certain solids (such as capacitors) need to
be preshredded prior to being fed into the kiln. Liquid wastes can be fed
directly into the afterburner section. The liquids are atomized using
compressed air, which produces a rotary action in the combustion zone.
    The combustor is a Loddby furnace measuring 1.6m diameter by 4.9m long.
The afterburner zone measures 4 x 4.3 x 10.6m. Natural gas and/or No. 2 fuel
oil are used as ignition fuel and also as a supplementary fuel if necessary.
Combustor temperatures can reach 1500°C, and afterburner temperatures average
1300°C. Residence times in the afterburner range from 2 to 3 seconds. Kiln
residence times vary widely according to the form of the waste, with residence
time being a function of design, solids content and viscosity. Combustion
gases from the afterburner pass through a combination venturi
scrubber/absorption tower system in which particulates and acid gases are
removed from the gas stream. A fraction of the scrubbing water is dosed with
lime and returned upstream of the venturi throat to increase scrubbing
efficiency. Induced draft fans are used to drive the scrubber gas stream to
the atmosphere.

ENSCO (M. M. Dillon, 1983; McCormick, 1986)--

    A schematic of the ENSCO incineration facility is shown in Figure 4.1.2.
Drummed wastes are fed to an enclosed shredder where solids drop into a hopper
and are conveyed by an auger into the rotary kiln. Liquid wastes are mixed
with the shredded solids and conveyed to the kiln or injected directly into
the combustion chamber. The air in the enclosed shredder is drawn by a fan
into the rotary kiln.
    The rotary kiln measures 2.1m in diameter by 10.4m long and is angled
slightly so that the solid residue flows by gravity to the ash drop. Flue
gases from the kiln are ducted to the 85 cubic meter combustion chamber where
fuel (often an organic waste) is burned to create a high temperature zone
(outlet temperature 1250°C). This afterburner, which possesses an outlet

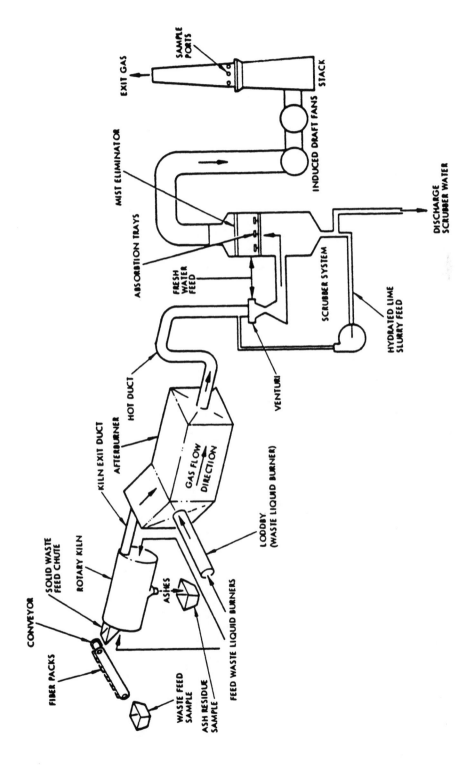

Figure 4.1.1. Schematic of Rollins Environmental Services' incinerator.
Source: Shih, C.C., et al., 1978.

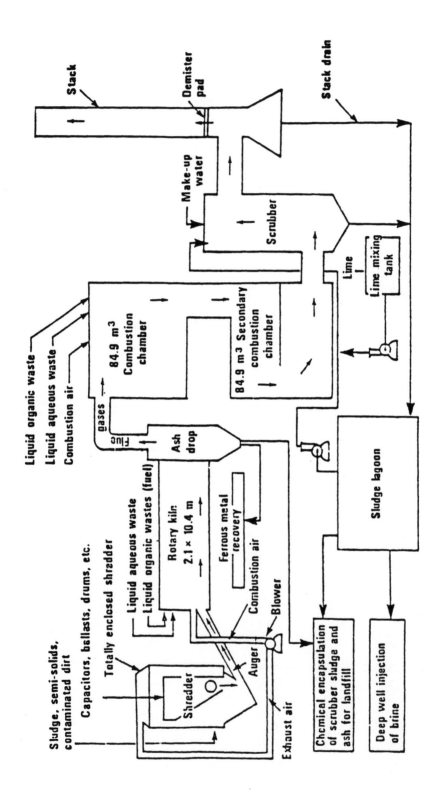

Figure 4.1.2. Schematic of ENSCO stationary incinerator [M.M. Dillon, 1983].

temperature of 1040°C, combusts the gaseous effluent from the primary combustor. Combustion products from the secondary combustion chamber are contacted with a caustic and lime slurry, passed through a spray tower scrubber, demisted, and finally discharged to the atmosphere. Kiln temperature, afterburner temperature, kiln and afterburner drafts, and carbon monoxide and carbon dioxide concentration are monitored regularly. Process instrumentation shuts down the feed for non-compliance with regulations.

The ENSCO incineration system can handle both solid and liquid wastes (the kiln section is not necessary for liquid wastes). Feed rates for PCB wastes are typically 8,140 kg/h with supplementary fuel requirements of 3,000 kg/h with an average of 16,400 kw heat input.

Restrictive Waste Characteristics--

The rotary kiln is capable of treating a wide variety of waste forms, including both liquids and solids, and also drums and bulk containers. It is capable of handling liquids and solids independently or in combination.

Spherical or cylindrical items may roll through the kiln before combustion is completed (i.e., insufficient residence time). Aqueous sludges may form clinker or ring residue on the refractory walls due to drying of the aqueous sludge wastes or melting of some solids.

Operating Parameters--

Typical operating parameters for stationary rotary kiln incinerators are summarized below (McGaughey, et al, 1984; M. M. Dillon, 1983; Bonner, 1981):

Residence Time	Ranges from a few seconds (highly combustible gas) to a few hours (low combustible solid waste).
Incinerator Temperatures	Can vary from 800 to 1600°C (1470 to 2900°F) depending on requirements for a particular waste.
Length to Diameter Ratio	Typically ranges between 2 and 10.
Peripheral Rotational Speed	Ranges from 1 to 5 fpm.
Incline Ratio	Ranges from 1/16 to 1/4 in./ft.

4.1.2   Technology Performance Evaluation

As discussed in the previous subsection, three commercial-scale
stationary rotary kiln incinerators have demonstrated six nines DRE (99.9999%)
for PCBs and are permitted to burn PCBs. Trial burns of dioxin wastes have
not been conducted at any of the commerical-scale stationary rotary kiln
incinerators due to strong public opposition. However, trial burns of
dioxin-containing wastes have been performed using the stationary rotary kiln
incinerator at the U.S. EPA Combustion Research Facility (CRF) in Jefferson,
Arkansas.

The CRF is a 3100 sq.ft. permitted experimental facility built for the
purpose of conducting pilot scale incineration burns and evaluating whether
incineration is an effective treatment/disposal option for various types of
hazardous waste. During September 1985, trial burns were conducted at the CRF
using dioxin-containing toluene still bottoms (Ross, et al., 1986). These
were generated by the Vertac Chemical Company in Jacksonville, Arkansas and
stored there pending an EPA decision on appropriate treatment/disposal.

The CRF contains two pilot-scale incinerators and associated waste
handling, emission control, process control, and safety equipment
(Carnes, 1984). Additionally, onsite laboratory facilities are available to
characterize the feed material and process performance samples. As shown in
Figure 4.1.3, the main components of the CRF incineration system include a
standard rotary kiln incinerator, an afterburner and a conventional air
pollution control system (Carnes, 1984; Ross, et al., 1986;
Ross, et al., 1984). Waste fed into the kiln flows countercurrent to the
primary burner (concurrent configuration is also possible). The
monitoring/control equipment for the kiln includes a propane meter, a pitot
tube to monitor primary combustion air, a shielded thermocouple to control
temperature, and monitoring equipment for combustion gas composition and flow
rate. Organic components are determined by extractive sampling through a
heat-traced sampling line and a liquid impinger (from EPA Method 5) or a
volatile organic sampling train (VOST) (Carnes, 1984; Ross, et al., 1986;
Ross, et al., 1984).

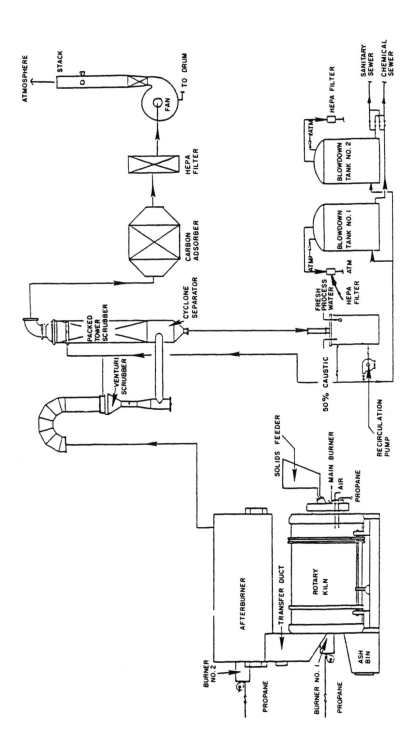

Figure 4.1.3. Simplified schematic of CRF rotary kiln (Ross, et al., 1986).

Following combustion in the kiln, the combustion gases are directed through a refractory-lined transfer duct to the afterburner. A shielded thermocouple is used to control temperature in the afterburner. Surface temperature, exit gas temperature and combustion gas composition and flow rate are monitored. Exhaust gases are cleaned in the air pollution control system which consists of a variable throat venturi scrubber, a fiberglass reinforced polyester wetted elbow, a packed tower caustic scrubber, and an induced draft fan (Carnes, 1984; Ross, et al., 1986; Ross, et al., 1984).

A total of four trial burns were performed in 1985 between September 4th and September 21st. These included a blank burn to establish background emission levels, a short-duration (4 hrs.) burn to establish feed capabilities and to test the sampling protocol, and two full waste burns (10 hr. duration) to establish the DREs for dioxin (Ross, et al., 1986).

The results of the two full waste burns are presented in Table 4.1.1. The data show that the 2,3,7,8-TCDD DRE was greater than 99.9997 percent as measured in the virtual stack (E-DUCT) which would correspond to the stack of an actual hazardous waste incinerator. The reason six nines DRE could not be established was that the detection limits experienced in the sampling and analysis protocols used were not sufficiently low (Ross, et al., 1986; Carnes, 1986). Despite this, it was concluded from the data in the study that "incineration under the conditions existing in the CFR pilot incineration system for these tests is capable of achieving 99.9999 percent dioxin DRE" (Ross, et al., 1986). It was further concluded that land-based incineration should be considered a viable disposal method for the Vertac still bottoms waste given that appropriate safeguards are employed (Ross, R. W., et al., 1986).

The concentrations of 2,3,7,8-TCDD and other CDDs and CDFs were also measured in the scrubber blowdown water and kiln ash (Ross, et al., 1986). The maximum concentration of 2,3,7,8-TCDD detected in the scrubber blowdown was 0.12 picograms per milliliter (approximately 0.1 ppt). In most samples of blowdown, all forms of CDDs were undetected at detection limits of 0.006 to 0.020 pg/ml. In one sample, however, 0.78 pg/ml of Octa-CDD was detected. No 2,3,7,8-TCDF was detected in any blowdown samples. Total TCDF was detected at 0.20 pg/ml in all 4 samples. Tetra-, penta-, hexa- and hepta- CDD and CDFs were not detected in any of the kiln ash samples at detection limits ranging from 1.3 to 37 picograms per gram (ppt).

TABLE 4.1.1. VERTAC STILL BOTTOM TEST BURN

Parameter	Test 1	Test 2
Mean Waste Feedrate (lb/hr)	22	39
Duration of Burn (hrs)	12	6
2,3,7,8-TCDD in Feed (ng/g)	37	37
Kiln Exit Gas Temp. (°C)	980	990
Nominal Kiln Residence Time (sec)	4.9	6.0
Afterburner Exit Gas Temp. (°C)	2030	2030
Afterburner Nominal Residence Time (sec)	1.8	2.3
DRE at Kiln Exit (%)[a,b]	>99.99995	>99.99976
DRE at E-Duct (%)[c,d]	>99.9997	>99.9997
Particulate at Stack (mg/dscm)	47.2[e]	255[f]
Maximum 2,3,7,8-TCDD Concentration in Scrubber Blow Down (pg/l)	0.12	
Maximum 2,3,7,8-TCDD Concentration in Ash (pg/g)	ND[g]	

[a]Based on Helium tracer data

[b]Average of values from two sampling trains

[c]Average values from 4 sampling trains

[d]Based on flue gas velocity data

[e]Uncorrectted

[f]Corrected to 7% oxygen

[g]Detection limits ranged from 5.6 to 28 pg/g

Reference: Ross, et. al., 1986.

Since in all cases the residues from this incinerator contained CDDs and CDFs at levels below 1 ppb, it would be expected that these residues could be land disposed in accordance with the screening levels proposed in 51 FR 1602. These screening levels are based on the use of a different analytical method (Method 8280) than used in the present situation. Therefore, definite conclusions cannot be made. Nonetheless, the concentrations of CDDs and CDFs detected in the treatment residues indicate that a high degree of destruction did occur.

Several problems encountered during the trial burns should be mentioned. These include (Ross, et al., 1986):

1. Waste was fed into the kiln through a water-cooled feed lance using a Moyno cavitation pump. The lance frequently became clogged due to carbon-buildup from coking of the waste material.

2. The test plan called for continuous monitoring of flue gas, CO_2, O_2, CO and NO_x, at the stack with one set of emission analyzers and at the kiln exit and afterburner exit on a time-share basis with another set. However, only one set of emission analyses was operational during most of the test series. Therefore, no kiln emission monitoring data were obtained. Very little simultaneous afterburner exit and stack data were obtained.

3. An air leak in the sample transfer line from the afterburner exit to the monitors caused the data to be "substantially compromised". Also, difficulties were encountered while monitoring by means of the isokinetic Method 5 sampling train (MM5). The glass frit in the MM5 train condensor/XAD2 sorbent cartridge frequently became plugged. Most exhaust and stack sampling was at less than 50 percent isokinetic, which compromises the particulate emission results.

4.1.3 Costs of Treatment

Currently, there are no stationary kilns permitted to burn dioxin wastes. Thus, no costs are available. However, these costs would be expected to be similar to or greater than the costs for PCBs incineration. Table 4.1.2 lists the average unit costs for PCBs wastes at the currently permitted stationary rotary kiln facilities.

TABLE 4.1.2.   AVERAGE UNIT COSTS FOR PCB WASTE DESTRUCTION AT
PERMITTED STATIONARY ROTARY KILN FACILITIES

PCB concentration range	Unit costs liquids	Unit costs solids
0 to 50 ppm	0.25 $/lb	0.40 $/lb
50 to 1000 ppm	0.30 $/lb	0.45 $/lb
1000 to 10,000 ppm	0.35 $/lb	0.50 $/lb
10,000 to 100,000 ppm	0.40 $/lb	0.60 $/lb
100,000 ppm	0.45 $/lb	0.70 $/lb

References: Bailey, May 1986; SCA, May 1986.

4.1.4 Process Status

Land-based incineration systems with potential to treat dioxin wastes include commercial incineration facilities which have been approved for PCB disposal, in addition to RCRA hazardous wastes. These incinerators are operated by Rollins Environmental Services (Deer Park, Texas), ENSCO (El Dorado, Arkansas) and SCA (Chicago, Illinois). Each of these systems contains a rotary kiln incinerator followed by an afterburner section which can also be used independently as a liquid injection incinerator.

The Rollins and ENSCO facilities can accept both liquid and solid wastes, but the SCA incinerator has only been approved for the disposal of liquid PCB wastes. The following are the maximum feed rates for these land-based incineration systems (GCA, 1985; Clarke, 1986):

Rollins (Deer Park, TX).	1,440 lb/hr for solids 6,600 lb/hr for liquids
ENSCO (El Dorado, AK.)	2,500 lb/hr for solids 5,000 lb/hr for liquids
SCA (Chicago, IL.)	2,910 lbs/hr for solids 6,300 lbs/hr for liquids

Although none of these facilities has conducted trial burns for the destruction of dioxin-contaminated waste, their ability to demonstrate six nines DRE for PCBs suggests that they would be able to destroy dioxins. Results of trial burns using the CRF pilot-scale rotary kiln incinerator show the potential for dioxin destruction at the RCRA regulated DRE.

4.2 MOBILE ROTARY KILN

As mentioned previously, the EPA Mobile Incineration System (MIS) has demonstrated six nines DRE on waste containing 2,3,7,8-TCDD. ENSCO has developed a modified version of the EPA mobile rotary kiln incinerator with improved solids handling techniques that is also capable of treating these wastes. These two units are described below.

4.2.1   Process Description

The EPA mobile unit is mounted on four heavy-duty semi-trailers which can be transported to a treatment site and connected in series. The system includes a rotary kiln, a secondary combustion chamber, and a scrubber, along with the following support equipment: bulk fuel storage, waste blending and feed equipment for both liquids and solids, scrubber solution feed equipment, ash receiving drums, stack monitoring equipment, and an auxilliary diesel power generator (IT Corporation, 1985a; Freestone, et al., 1985; Freeman and Olexsey, 1986). A schematic of the EPA mobile incineration system is shown in Figure 4.2.1.

The first trailer consists of a solids feed system, burners, and a rotary kiln (IT Corporation, 1985a; GCA, 1985; Freeman and Olexsey, 1986; Freestone, et al., 1985). After being shredded by the shear-type shredder (driven by a 150-HP motor), the solids are carried by an enclosed conveyor to the hopper. A sliding knife gate at the bottom of the hopper opens to allow the contents of the hopper to empty into the ram feeder. The hydraulically operated ram feeder forces the feed into the kiln. The kiln is rated at 15 million BTU/hr and is capable of handling 150 lbs/min of dry solids. Contaminated water and contaminated fuel oil are introduced directly into the afterburner at maximum feed rates of 3 gpm and 1.7 gpm, respectively. The refractory lining material in the kiln is abrasion-resistant, acid-resistant, and able to withstand repeated thermal cycling. The rotary kiln operates at approximately 1800°F to vaporize and partially combust the incoming waste. Incombustible ash is discharged directly from the kiln. Ducting connects the solids feed system and the ash discharge system to a trailer-mounted, dual-manifold, high efficiency particulate air (HEPA)/carbon filter system to control fugitive dust emissions.

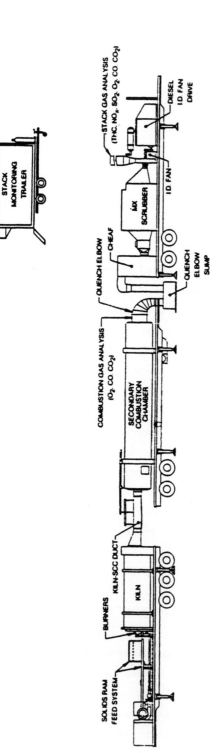

Figure 4.2.1.   Schematic of EPA mobile incineration system [U.S. EPA, 1984].

The exit gases from the kiln pass through a secondary combustion chamber (SCC) located on the second trailer. The secondary combustion chamber is capable of providing 2.2 seconds residence time at 2200°F (1200°C) to complete the combustion process. The refractory lining of the SCC is coated with a thin layer of acid-resistant mortar. The flue gases from the SCC are cooled by water sprays to 190°F. Excess water is collected in a sump.

The cooled gases are then passed through air pollution control equipment. The third trailer contains the air pollution control system which consists of a quench system with partial acid gas removal, a clean high efficiency air filter (CHEAF) to remove submicron-sized particulates, and flue gas instrumentation. A draft fan moves combustion gases through the system and maintains a negative pressure so that toxics will not escape to the atmosphere.

The cleaned gases are emitted from the system through a 40 ft high stack. Combustion and stack gas monitoring equipment are contained within the fourth trailer. Carbon dioxide and oxygen are continuously monitored in the stack.

The ENSCO mobile incinerator is similar in design to the EPA mobile incinerator. The ENSCO mobile system is capable of incinerating 150 gal/hr of liquid PCB waste blend or 2 to 5 tons/hr of solid hazardous waste (Pyrotech Systems, 1985). The system is mounted on trailers and consists of the following six basic process modules: solids incineration, liquids incineration, waste heat boiler, a pollution control/prime mover system, control room and process laboratory, and effluent neutralization and concentration systems (Pyrotech Systems, 1985; Sickels, 1986). A schematic of the process is shown in Figure 4.2.2.

The solids incineration module consists of a rotary kiln, solids preparation and charging equipment, a burner, an air blower, and an ash discharge system (Pyrotech Systems, 1985; Sickels, 1986). During operation, contaminated solids are shredded and fed to the kiln via a ram feeder. The rotary kiln is a brick-lined, carbon steel pipe measuring 30 ft in length with an inside diameter of 69 inches. It is mounted horizontally with a slight vertical incline, and rotated by a trunnion drive mechanism. The kiln has a nominal solids capacity of 3.5 tons/hr and a nominal heat load of 15 million Btu/hr. (Pyrotech Systems, 1985; Sickels, 1986).

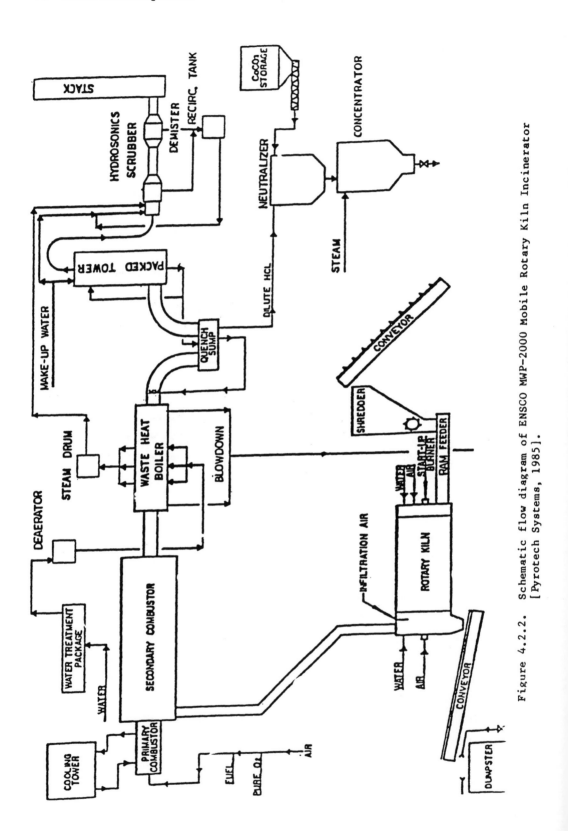

Figure 4.2.2.   Schematic flow diagram of ENSCO MWP-2000 Mobile Rotary Kiln Incinerator [Pyrotech Systems, 1985].

The liquid incinerator/afterburner (SCC) is a large refractory-lined vessel designed to maintain a temperature greater than 2200°F (1200°C) and a residence time greater than 2 seconds. It has a length of 40 ft and an inside diameter of 7 ft. The nominal heat load for the SCC is 20 million Btu/hr. Combustion can be enhanced by oxygen enrichment (Pyrotech Systems, 1985; Sickels, 1986).

A firetube boiler is used to recover heat in the offgas stream and to provide the steam required to drive the system prime mover. A scrubber/ejector system acts as the system prime mover, and also as the primary pollution control device. Thus, an induced draft fan is not required. Offgases containing particulates are driven through the ejector nozzle where they are contacted with water. Turbulent mixing occurs causing efficient particulate capture (emissions less than 0.08 gr/DSCF*). The agglomerated particulate and water are subsequently removed by a horizontal gas water separator (HGW) with an integral demister. Cleaned gases are vented to the atmosphere. Scrubber water is filtered to remove solids and then recycled to the steam ejector. The combustion processes and emissions control systems are monitored and controlled by an automated system. Fail-safe shut-down occurs when the following events are detected: loss of combustion flame, low flue pressure, secondary combustor pressurization, or excessive temperature at the control device inlet (Pyrotech Systems, 1985; Sickels, 1986).

Restrictive Waste Characteristics--

The EPA and ENSCO mobile incinerators are able to handle both solid and liquid hazardous waste streams. Bulky solids need to be fed through a specially designed solids feed system (described previously) prior to being fed to the kiln. Wastes with a low heating value (i.e., less than 8,000 Btu/lb) may require blending with kerosene prior to being fed to the combustor (GCA, 1984).

---

* gr/DSCF = grains per dry standard cubic foot.

Operating Parameters--

Operating parameters for the two mobile rotary kiln systems are summarized below (U. S. EPA, 1984; Sickels, 1986; Freestone, et al., 1985):

	EPA/MIS	ENSCO
Waste Forms	Solids, Liquids	Solids, Liquids
Maximum Waste Feed Rate (lb/hr)		
-Solids to Rotary Kiln	9,000	10,000
-Liquids to Rotary Kiln		3,000
-Liquids to SCC	1,500	4,000
Kiln Temperature (°F)	1800	1800
SCC Temperature (°F)	2200	2200
SCC Residence Time (sec)	2.2	2

Severe weather conditions can effect the operation of the mobile incinerator. For instance, extremely cold weather during the initial stages of the EPA Mobile incinerator trial burns on the Denney Farm site caused the No. 2 diesel fuel to gel, hydraulic fluids to thicken, and water lines to freeze (IT Corporation, 1985a; Krogh, 1985).

4.2.2   Technology Performance Evaluation

Initial trial burns with the EPA mobile incinerator were conducted in Edison, New Jersey using surrogate compounds to mimic RCRA-listed constituents such as dichlorobenzene, trichlorobenzene, tetrachlorobenzene, and PCBs. In these liquid waste trial burns, up to six nines DRE of PCBs was demonstrated. Following this, laboratory studies were conducted to establish optimum conditions for treating soils contaminated with dioxin. The following conclusions were made based on these studies (IT Corporation, 1985a):

- Thermal treatment of contaminated Missouri soils was capable of achieving 1 ppb or lower concentration of residual 2,3,7,8-TCDD and other chlorinated dioxins and chlorinated furans in the incinerator ash.

- Temperature and time are the primary factors which affect 2,3,7,8-TCDD removal efficiency.

- Other factors (such as initial moisture content, soil type, etc.) had either no effect or a minor effect on dioxin removal. An exception was very large soil agglomerates with high initial moisture content, which require substantially longer residence time to achieve a uniform temperature throughout the solid mass.

In December 1984, a solids feed system (see process description) was installed so that the MIS could be tested on dioxin-contaminated soils at the Denney Farm Site in Missouri. Prior to testing at the Denney Farm Site, a series of "shakedown" tests were performed to determine the conditions necessary to decontaminate the soils (IT Corporation, 1985a).

Four different types of soils were passed through the solids feed system, rotary kiln, and ash discharge system without showing any problems which adversely affected performance (IT Corporation, 1985a; U. S. EPA, 1985). However, some slagging occured as solids were carried over into the front end of the SCC where they accumulated to a depth of up to 9 inches. Residence times averaged over 30 minutes with a 2000 lb/hr feed rate. The kiln rotated at 1 rpm and discharged ash at an average temperature of 750°C. Stack particulate tests demonstrated that particulate emissions were well below RCRA limits (180 mg/m^3, corrected to 7 percent oxygen). These data are summarized in Table 4.2.1.

After completing the shakedown tests, the MIS was transported to the Denney Farm Site in Missouri for a series of dioxin trial burns (U. S. EPA, 1985; U. S. EPA, 1984). As indicated in Table 4.2.2, greater than 99.9999 percent DRE was achieved on waste feeds of up to 2000 lbs/hr of solids and 250 lbs/hr of liquids (IT Corporation, 1985a). The concentration of 2,3,7,8-TCDD in liquid waste feed materials ranged from 225 to 357 ppm, and in the solid wastes from 101 to 1010 ppb (IT Corporation, 1985a and b; U. S. EPA, 1985).

In addition to the 99.9999 percent DRE, the treated soil and process wastewater from these burns were analyzed for a number of constituents and were shown to meet the delisting guidelines (see Table 4.2.3) established by the Office of Solid Waste (Poppiti, 1985).

TABLE 4.2.1.   SOILS USED IN THE EPA MOBILE INCINERATOR DURING
PRELIMINARY TESTING OF THE SOLIDS FEED SYSTEM

Soil type	Test purpose	Particulate emissions
Denney Farm Area Soil	To ensure that there were no unusual problems with soil from that area; Site soil was very dry from being stored	18 mg/Nm^3
Montmorillonite	Planned to be used as the Solids Carrier in Test 1 of Solids Trial Burn	17 mg/Nm^3
Coral from Florida	Potential Future Use of EPA/MIS for U.S. Air Force on Johnston Island Contaminated Coral	9 mg/Nm^3
Clarksburg (New Jersey) Soil	Readily Available Soil in the Missouri Area	Not Available

Reference: IT Corporation, 1985a.

TABLE 4.2.2.   RESULTS OF DIOXIN TRIAL BURNS IN MISSOURI
USING THE EPA MOBILE INCINERATOR SYSTEM

Parameter	Permit limit	Test number 2-2[a]	2-3	2-4	2-5	3-1,-2,-3[b]
Rotary kiln						
Temp. (°C)	760-1040	845	826	902	918	852
Secondary						
Temp. (°C)	1120-1315	1153	1199	1194	1199	1187
Ret. time (sec)		2.5	3.2	2.5	2.6	---
O_2 (vol.%)	4	7.96	6.61	6.44	6.42	5.4
CO (ppm_v)	100 (6 min)	7.7	1.3	2.3	2.5	6.8
CO_2 (vol.%)		10.9	11.3	11.4	11.1	---
C.E. (%)		99.993	99.999	99.998	99.998	---
TH (ppm_v)		0.1	0.5	0.5	0.9	---
NO_x (ppm_v)		139	132	126	166	---
Liquid waste feed						
Total flow (#/hr)		233	236	234	247	---
TCDD (ppm)	400	249	357	264	225	---
TCDD (g/hr)	27.2	26.3	38.3	28.0	25.2	---
Solids feed						
Total flow (#/hr)	2000	1158	1163	2068	1322	973
TCDD (ppb)		101	382	1010	770	---
TCDD (g/hr)		0.05	0.24	0.95	0.46	---
Stack						
Emission Rate:						
TCDD[c](mg/day)		0.169	0.127	0.031	0.064	---
Particulates (mg/Nm^3@7% O_2)	180	134.3	147.3	145.6	201.5	---
DRE (%)	99.9999	99.99997	99.99998	99.99999	99.99998	---

TCDD = 2,3,7,8-TCDD; C.E. = combustion efficiency;
DRE = destruction and removal efficiency; TH = total hydrocarbons.

[a]Test 2   liquid feed: dioxin-contaminated TCP still bottoms
solids feed: TCP still bottoms and contaminated soil.

[b]Test 3   liquid feed: fuel oil; solids feed: dioxin-contaminated/brominated
naphthalene-contaminated lagoon sludge; combined results for three
one-half hour runs.

[c]No TCDD detected at detection limits in effect; hence, emission rates denoted
are detection limits and not measured quantities.

References: IT Corporation, 1985a; IT Corporation, 1985b

TABLE 4.2.3.   MISSOURI DEPARTMENT OF NATURAL RESOURCES AND EPA DELISTING PARAMETERS FOR ORGANIC CONSTITUENTS IN INCINERATOR ASH AND SCRUBBER WASH WATER

Toxic Constituent	Concentration	
	Ash	Scrubber water
Dioxins/Dibenzofurans[a]	1 ppb	10 ppt
2,3,4-Trichlorophenol	100 ppm	10 ppm
2,3,5-Trichlorophenol	100 ppm	10 ppm
2,4,6-Trichlorophenol	1 ppm	50 ppb
2,5-Dichlorophenol	350 ppb	15 ppb
3,4-Dichlorophenol	100 ppm	10 ppm
2,3,4,5-Tetrachlorophenol	1 ppm	50 ppb
2,3,4,6-Tetrachlorophenol	1 ppm	50 ppb
1,2,4,5-Tetrachlorobenzene	100 ppm	10 ppm
1,2,3,5-Tetrachlorobenzene	100 ppm	10 ppm
Hexachlorophene	200 ppm	5 ppm
Polychlorinated Biphenyls	2 ppm	1 ppm
Benzo(a)pyrene	5 ppm	10 ppb
Benzo(a)anthracene	5 ppm	10 ppb
Chrysene	50 ppm	1 ppm
Dibenzo(a,h)anthracene	5 ppm	10 ppb
Indeno(1,2,3-c,d)pyrene	5 ppm	10 ppb
Benzo(b)flouranthene	5 ppm	10 ppb

[a]Weighted average of TCDDs/TCDFs, PeDDs/PeDFs, and HxCDDs/HxCDFs using toxicity weighting factors.

Reference: Poppiti, 1985; U.S. EPA, 1985.

TABLE 4.2.4.   MATERIAL TO BE INCINERATED DURING FIELD
DEMONSTRATION OF THE EPA MOBILE INCINERATOR SYSTEM

Material	Estimated quantity	2,3,7,8-TCDD Conc.
*Denney Farm		
Soil	210 cubic yards	500 ppb
Mixed solvents and water	2,590 gal	Low
Chemical solids and soils	31,150 lb	1 ppb - 2 ppm
Drum remnants and trash	84 85-gal overpack drums	Unknown
Verona		
Hexane/Isopropanol	10,000 gal	0.2 ppm
Methanol	5,000 gal	ppt
Extracted still bottoms	5,000 gal	0.2 ppm
Activated carbon	5,000 lb	Unknown
Decontamination solvents	1,000 gal	Unknown
Sodium sulfate salt cake	23 cubic yards	1 ppb
Miscellaneous trash	84 55-gal drums	Unknown
*Neosho		
Spill area soil	25 cubic yards	60 ppb
Bunker soil/residue	15 drums	2 ppm
Tank asphaltic material	75 gal	2 ppm
*Erwin Farm		
Empty drums with residue	30 drums	8 ppb
*Rusha Farm		
Spill area soil	10 cubic yards	Unknown
*Talley Farm		
Spill area soil	10 cubic yards	6 ppb
Eastern Missouri		
Times Beach soil sample	3 cubic yards	500 ppb
Piazza Road soil sample	3 cubic yards	1,600 ppb

*These burns have been completed, but the results were not released in time to
be included in this document.

References: Krogh, 1985; IT Corporation, 1985a

Average operating parameters during the trial burn for dioxin-contaminated soil and liquids were as follows (IT Corporation, 1985 a and b; U. S. EPA, 1985):

Kiln Temperature	1800°F
SCC Temperature	2200°F
SCC Combustion Gas Flow Rate	13,500 acfm
SCC Residence Time	2.6 seconds
Waste Feed Rate	2000 lb/hr (soil) 250 lb/hr (liquid)

Auxiliary Fuel
   -Kiln   5 to 6 million Btu/hr
   -SCC   4 to 5 million Btu/hr

Following the successful (demonstrating 99.9999 percentDRE) completion of these preliminary trial burns, additional test burns were planned for the EPA/MIS as summarized in Table 4.2.4. As noted in the table, burns of the material from five of these sites (Denney Farm, Neosho, Erwin Farm, and Talley Farm) have been completed. The total amount of dioxin-contaminated material that was successfully burned (i.e., achieved greater than six nines DRE) included 2 million pounds of soils and 180,000 pounds of liquids (Hazel, 1986; Freestone, 1986). The material from the remaining sites listed in Table 4.2.4 is scheduled to be burned as soon as funding becomes available (Hazel, 1986). Currently, the amount of dioxin-contaminated material that remains to be burned includes 600,000 pounds of soil and 80,000 pounds of liquid wastes (Hazel, 1986).

During the incineration of the dioxin-contaminated wastes from these sites, several parameters were monitored continuously, including: CO, CO_2, O_2, NO_x, operating temperatures and feed rates (Hazel, 1986; Freestone, 1986). Built-in safety controls cause the operations to stop if any of these parameters are not within the proper range (Hazel, 1986). In addition, the waste residues from the burns (i.e., ash and water) are continually monitored. To date, dioxin has never been found in the burn residues (Hazel, 1986). More detailed data will be available in the Final Report scheduled for release within the next few months (Freestone, 1986).

ENSCO has constructed four mobile units; two MWP-2000 type incinerators and two MWP-75 incinerators (McCormick, 1986). The MWP-75 (earlier model) was tested on PCB-containing wastes in 1983 (Freestone, et al., 1985; GCA, 1985). The results of these tests, listed in Table 4.2.5, demonstrated successful (>99.9999 percent DRE) destruction of the PCBs (IT Corporation, 1985a). One of the MWP-2000 units, currently located at a site near Tampa, Florida, is undergoing test burns on PCB-containing soils (McCormick, 1986). Initially, ENSCO encountered slagging problems similar to those experienced with the EPA mobile incinerator (IT Corporation, 1985a; Lee, 1985). To solve these problems, six chutes were added to the secondary unit and a cyclone was installed prior to the secondary unit to remove fine particulates (IT Corporation, 1985a). The second MWP-2000 unit was used recently to conduct a series of trial burns at the El Dorado, Arkansas facility using dioxin-containing wastes from the Vertac Site (Hicks, 1986). The purpose of these burns was to obtain RCRA certification. The results of the dioxin trial burns are expected to be released in late May or early June, 1986 (McCormick, 1986).

4.2.3   Costs of Treatment

A detailed procedure for estimating the unit costs of a mobile rotary kiln incinerator has not yet been developed. The EPA intends to develop such a procedure and present it in its final report on the Denney Farm Trial Burns (IT Corporation, 1985a).

TABLE 4.2.5.   EMISSION DATA FOR THE ENSCO MOBILE INCINERATOR (MWP-75)
PCB TRIAL BURN

Condition	Result
PCB DRE	>99.9999%
Carbon Monoxide	20 ppm
Nitrogen Oxides	300 to 500 ppm
Particulate	Met or exceeded all standards
HCl Scrubbing	99% at 1,500 lb HCl/hr

Reference: Sickels, 1986; Pyrotech Systems, 1985.

Factors which need to be considered when developing a cost estimate for a
mobile incineration system are as follows (IT Corporation, 1985a).

- Labor is a substantial portion of the operating costs. The
  greater the capacity of the system (e.g., by size or number of
  systems or through design modifications), the lower the unit
  costs.

- Costs are lower if the heat and moisture content of the waste
  feed are low because of the increased feed rate that can be
  maintained (up to 200 lb/hr).

- The greater percentage of time that the system is able to process
  material, the lower the overall costs.

- Setup costs vary according to site design requirements, permit
  costs, etc.

- The longer the duration of the cleanup operations, the higher the
  costs will be.

- Reliability (via conservative design and redundancy) increases
  actual operating time and thereby reduces overall costs.

4.2.4 Process Status

The EPA mobile incinerator was developed and tested for the purpose of
evaluating the technical and economical feasibility of mobile incineration, to
establish procedures for obtaining Federal, State and Local permits and to
gauge public reactions (IT Corporation, 1985a; Krogh, 1985). The intention of
future uses of the EPA mobile incinerator is to encourage commercial
development of onsite cleanup technologies. As a result of conducting
research burns on various waste forms, the EPA is able to provide operating
specifications and other valuable information to the private sector. The EPA
believes that, given this information, the private sector will.be capable of
developing improved, more reliable, larger capacity, lower cost systems.

ENSCO has developed a modified version of the EPA mobile rotary kiln
incinerator, with improved solids handling techniques (Pyrotech Systems,
1985). ENSCO has three mobile rotary kiln units (termed the MWP-200 series),
which are designed to process 3 to 4 cubic yards/hour (Hicks, 1986; McCormick,
1986). The MWP-2000 incinerators are capable of burning both solids and
liquids using supplimentary fuel oil for wastes with low BTU content.

Initially, ENSCO encountered slagging problems which were solved by adding six chutes to the secondary unit and a cyclone was installed prior to the secondary unit to remove fine particulates (Pyrotech Systems, 1985). These modifications increase the treatment costs.

One of the MWP-2000 units is currently located at a site near Tampa, Florida where it is being used to clean up a site containing liquids contaminated with chlorinated organics (McCormick, 1986; Lee, 1985). A second MWP-2000 unit located at the El Dorado, Arkansas facility (i.e., the location of the ENSCO stationary rotary kiln system) has just completed a series of dioxin trial burns using wastes from the Vertac Site. The results are expected to be released in May 1986 (McCormick, 1986). The construction of the third MWP-2000, a computer-operated unit, is not complete yet. Upon completion, this third unit is scheduled to undergo tests by the Air Force to handle dioxin-contaminated coral at Johnston Island (Lee, 1985; McCormick, 1986).

4.3   LIQUID INJECTION INCINERATION

4.3.1   Process Description

The general components of a liquid injection (LI) incineration system include a burner, primary combustion chamber, secondary (unfired) combustion chamber, quench chamber, scrubber, and stack. The LI incinerator system can be configured either vertically or horizontally. With the vertical configuration shown in Figure 4.3.1, the incinerator acts as its own stack and a portion of the stack may serve as a secondary combustion chamber. Vertical units are preferred for wastes which are high in inorganic salts and ash contents. In contrast, horizontal incinerators are connected to tall stacks, and may be used with low-ash wastes (Bonner, 1981; McGaughey, et al., 1984).

To ensure efficient combustion, the liquid must be atomized prior to entering the combustor. Atomization is typically accomplished either mechanically through rotary cup or pressure atomization systems, or via gas/fluid nozzles using high pressure air or steam. The distribution of spray (by volume) is more uniform with rotary cups than with pressure or air-atomized nozzles. Waste feed storage and blending tanks aid in maintaining a steady, homogeneous waste flow. Particle size in slurries is a critical factor for successful operation because the burners are susceptible to clogging by particulate or caked material at the nozzles (McGaughey, et al., 1984; Bonner, 1981).

Combustion chamber residence times generally range from 0.5 to 2.0 seconds. Operating temperatures depend on the waste type and destruction requirements, but typically range from 650 to 1750°C (1200 to 3180°F) (McGaughey, et al., 1984; GCA, 1985). The heat capacity (BTU) of the waste liquid must be adequate for ignition and incineration or a supplemental fuel must be added.

Modified LI incineration systems which have been used to destroy dioxin wastes are employed in ocean incinerators. The Vulcanus uses a vertically configured system. Two identical, refractory-lined furnaces are located at the stern. Each incinerator consists of a combustion chamber and a stack.

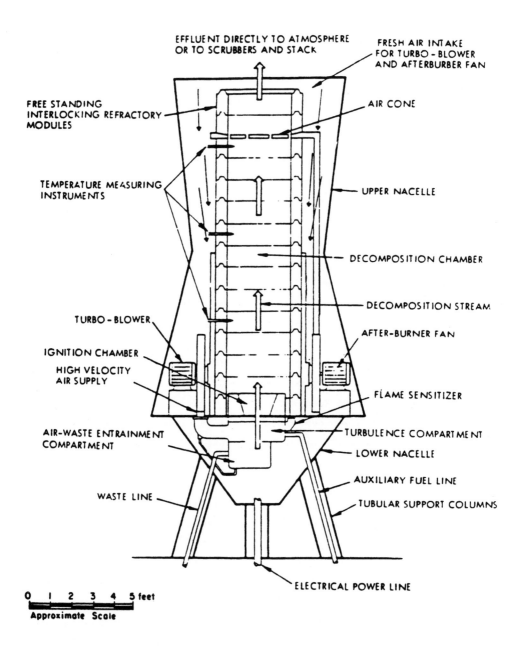

EFFLUENT DIRECTLY TO ATMOSPHERE
OR TO SCRUBBERS AND STACK

FRESH AIR INTAKE
FOR TURBO - BLOWER
AND AFTERBURBER FAN

FREE STANDING
INTERLOCKING REFRACTORY
MODULES

AIR CONE

TEMPERATURE MEASURING
INSTRUMENTS

UPPER NACELLE

DECOMPOSITION CHAMBER

DECOMPOSITION STREAM

TURBO - BLOWER

AFTER-BURNER FAN

IGNITION CHAMBER

HIGH VELOCITY
AIR SUPPLY

FLAME SENSITIZER

AIR-WASTE ENTRAINMENT
COMPARTMENT

TURBULENCE COMPARTMENT

LOWER NACELLE

WASTE LINE

AUXILIARY FUEL LINE

TUBULAR SUPPORT COLUMNS

ELECTRICAL POWER LINE

0   1   2   3   4   5 feet
Approximate Scale

Figure 4.3.1.   Vertically-oriented Liquid Injection Incinerator
                [Bonner, 1981].

Electrically driven pumps are used to deliver the waste feed to the combustion system. A gorator (mixing device) is used to reduce solids in the waste to a pumpable slurry (U. S. EPA, 1983; Ackerman, 1986; Ackerman, 1983). It also functions as a means of mixing the waste contents by recirculating the waste through the waste tank (U. S. EPA, 1983; U. S. EPA, 1978; Ackerman, 1986; Ackerman, 1983).

Each incinerator is equipped at the base with three rotary cup type burners. Waste or fuel oil is delivered through a central tube to atomization nozzles at the periphery of the rotating cup. Optimal mixing of the combustion gases is accomplished by positioning the burners tangentially to the vertical axis of the incinerators. Waste oil and fuel oil cannot be fed into a burner simultaneously but alternate burners can be operated with fuel and waste to alter required combustion temperatures.

Expected emissions from the stack include CO_2, CO, HCl, and H_2O vapor. Land-based liquid injection incineration systems require the use of a scrubber to remove acids prior to releasing combustion gases to the atmosphere. Ocean incinerators do not require scrubbers because the acids are expected to be neutralized by the ocean (U. S. EPA, 1983; Ackerman, 1986; Ackerman, 1983).

Restrictive Waste Characteristics--

Liquid injection incineration can only be used to dispose of combustible liquids or slurries which have a low enough viscosity to be pumped (i.e., less than 10,000 Standard Saybolt Units (SSU)) (GCA, 1985; U. S. EPA, 1983; Bonner, 1981). High viscosity also impairs atomization which can result in lower DREs. Liquid injection is most effective when the waste is physically and chemically homogeneous.

Particle size in slurries is a critical factor for sustained operation because the burner nozzles are susceptible to clogging by particulate or caked material (U. S. EPA, 1983; U. S. EPA, 1978; Ackerman, 1986). A gorator can be used to reduce solids in the waste to a pumpable slurry.

The heat capacity (Btu) of the waste liquid must be adequate for ignition and incineration, or a supplimental fuel must be added. If the heat content of waste is less than 8,000 Btu/lb, then a supplemental fuel is required.

Operating Parameters--

Typical operating parameters for the vertically-configured LI incinerator on the Vulcanus are as follows (U. S. EPA, 1983; U. S. EPA, 1978):

Residence Time	0.5 to 2.0 seconds
Temperature	650 to 1750°C (1200 to 3180°F)
Air Feed Rate	65,000 to 75,000 m^3/hr
Waste Feed Rate Capacity	7 to 10 tons/hr

4.3.2   Technology Performance Evaluation

The only documented burns of waste containing dioxins in a liquid injection incinerator are those that took place on board incinerator ships. These burns took place between July and September 1977. Three shiploads (totalling approximately 10,400 metric tons) of U.S. Air Force stocks of Herbicide Orange were incinerated by the M/T Vulcanus in the Pacific Ocean west of Johnston Atoll (U. S. EPA, 1978). A summary of the DREs achieved in these burns and other U.S.-sponsored ocean burns is presented in Table 4.3.1. Operating parameters for these Herbicide Orange trial burns are summarized in Table 4.3.2.

The Herbicide Orange stock consisted of an approximately 50-50 volume mixture of the n-butyl esters of 2,4-dichlorophenoxyacetic acid (2,4-D) and 2,4,5-trichlorophenoxyacetic acid (2,4,5-T) (U. S. EPA, 1978). A small quantity of the stock contained a 50-50 volume mixture of 2,4-D and the iso-octyl ester of 2,4,5-T. Certain lots also contained 2,3,7,8-TCDD ranging in concentration from 0 to 47 ppm (with an average concentration of 1.9 ppm). Drums containing the waste stock and waste handling equipment were rinsed with diesel fuel which was subsequently mixed with the waste feed to increase its heating value for incineration (U. S. EPA, 1978).

TABLE 4.3.1. SUMMARY OF TEST RESULTS FOR U.S.-SPONSORED OCEAN
TRIAL BURNS USING LIQUID INJECTION INCINERATION

Date	Principal hazardous constituents	Average %-DRE
11/74	Bis(dichloroyl)ether	>99.99998
	Tetrachloroethane	>99.99994
	2-Chloroethyl formate	>99.9997
	1,1,2-Trichloroethane	99.997
	1,2,3,4-Tetrachlorobutane	>99.99992
	1,2,3-Trichloropropane	>99.992
	2,3-Dichloro-1-propanol	99.996
	1,1-Dichloroethane	>99.99992
	1,2-Dichloroethane	>99.99998
	Dichloropropene	>99.99997
	1,2-Dichloropropane	99.9995
8/77	2,4-D	>99.999
	2,4,5-T	>99.999
	TCDD	>99.93
8/82	PCB with 1 chlorine atom	>99.997
	PCB with 2 chlorine atoms	>99.9993
	PCB with 3 chlorine atoms	>99.9997
	PCB with 4 chlorine atoms	>99.9998
	PCB with 5 chlorine atoms	>99.99999
	PCB with 6 chlorine atoms	>99.99992
	PCB with 7 chlorine atoms	>99.99993
	PCB with 8 chlorine atoms	>99.99994
	Chlorobenzenes	
	with 1 chlorine atom	>99.9998
	with 2 chlorine atoms	>99.9994
	with 3 chlorine atoms	>99.99996
	with 4 chlorine atoms	>99.99994
	with 5 chlorine atoms	>99.9999
	with 6 chlorine atoms	>99.99992
2/83	1,1-Dichloroethane	99.99994
	1,2-Dichloroethane	99.99996
	1,1,2-Trichloroethane	>99.999995
	Chloroform	99.9996
	Carbon Tetrachloride	99.998

References: U.S. EPA, 1983; U.S. EPA, 1978; Lee, 1985.

TABLE 4.3.2.   SUMMARY OF OPERATING PARAMETERS FOR
HERBICIDE ORANGE TRIAL BURNS USING LIQUID
INJECTION INCINERATION ON THE VULCANUS

Flame Temperature	1375-1610°C
Furnace Wall Temp.	1100-1200°C
Residence Time	1.0 to 2.0 seconds

Reference: U. S. EPA, 1978

During the period from December 1981 through January 1982, the first
ocean burn of PCBs was performed in U.S. waters (U. S. EPA, 1983). A second
shipload of PCB-containing wastes were incinerated aboard the M/T Vulcanus
during August 15-31, 1982 (U. S. EPA, 1983). Operating parameters for the
PCBs burns are summarized in Table 4.3.3. An EPA-sponsored test project was
performed during the second PCBs trial burn to measure emissions of
polychlorinated biphenyls (PCBs), chlorobenzenes (CBs),
tetrachlorodibenzofurans (TCDFs), and tetrachlorodibenzo-p-dioxins (TCDDs).

TABLE 4.3.3   SUMMARY OF OPERATING PARAMETERS FOR
PCB TRIAL BURNS USING LIQUID INJECTION
INCINERATION ON THE VULCANUS

Flame Temperature	1648-2048°C
Furnace Wall Temp.	1281-1312°C
Residence Time	1.1 to 1.5 seconds
Feed Rate	5.23-6.79 mt/hr

Reference: U. S. EPA, 1983.

No TCDDs were detected in any sample of waste or stack gas during the
tests. Detection limits for the waste ranged from less than 2 ppb to less
than 22 ppb. TCDFs, however, were detected in the waste feed at ppb levels
and in the stack gas at low ppt levels. [USEPA, 1983]

Even though land-based liquid injection incinerators have not been tested using dioxin-containing wastes, several have demonstrated greater than 99.9999 percent DRE on PCB-wastes. One of these is General Electric's Thermal Oxidizer in Pittsfield, MA. This unit, built by the John Zink Company in 1972, is a horizontal unit with a primary combustion chamber and secondary quiescent reactor followed by a vertical quench chamber and a scrubber. In 1981, a PCB trial burn was conducted. The results of this trial burn are presented in Table 4.3.4. As indicated in the table, greater than 99.9999 percent DRE of PCB was achieved at a temperature of 1141 to 1262°C and a residence time of 4.02 seconds. In addition, no PCBs were detected in the scrubber water discharge (Thayer, J.H. et al., 1983).

4.3.3   Costs of Treatment

The costs of treatment will be a function largely of the type of waste being fed to the unit. Aqueous wastes with low Btu values will require auxiliary fuel and consequently the costs will be higher. Wastes with high halogen content will also be more costly to treat because a caustic scrubber is required to remove halogen acid gases. A typical liquids incineration cost for halogenated solvents containing significant amounts of water (i.e., greater than 50%) is $200/metric ton. A PCB contaminated oil, because of the six nines DRE requirement and because PCBs are a highly toxic material may cost $500 metric ton to incinerate. [CGA Corporation, 1984] Incineration of TCDD-contaminated liquids would probably have costs similar to the latter and not the former waste type.

4.3.4   Process Status

There have been no liquid injection incinerators that have demonstrated six nines DRE on wastes containing dioxin. There are, however, at least two liquid injection incinerators that have been permitted to burn PCB wastes. Both of these incinerators are owned and operated by General Electric. One is in Waterford, N.Y., but it is apparently not available for commercial use, and the other is in Pittsfield, MA (McInnes, R.C. and R.C. Adams 1984). In

TABLE 4.3.4.   SUMMARY OF LIQUID INJECTION INCINERATION TRIAL BURN
RESULTS FOR PCBs – GENERAL ELECTRIC, PITTSFIELD, MA

Parameter	Trial burn results
Temperature Inside the Reactor Chamber	1,262°C – 1,141°C (2,303°F – 2,085°F)
Residence (Dwell) Time of Combustion Products	4.02 sec
Combustion Efficiency	99.993%
Oxygen Concentration during PCB Incineration	9.5 – 10.5%
Waste Oil Firing Rate	1.09 – 119 GPM
PCB Concentration in Oil	18.4 – 20.0%
Average PCB Destruction Efficiency	99.999982%
Average PCB Destruction and Removal Efficiency	99.999982%
HCl Scrubber Efficiency	99.82%
Particulate Emissions @ 12% CO_2	0.543 lb/hr 0.0361 gr/dscf
NO_x Emissions	18.3 ppm 0.43 lb/hr
RCl Emissions	0.000304 ppm 0.00002542 lb/hr
HCl Emissions	0.2752 lb/hr

Reference: Thayer, et al., 1983.

addition, Occidental Chemical Company is reportedly in the process of trying
to get a permit to perform test burns of non-aqueous phase leachate (NAPL)
from the Hyde Park landfill in its liquid injection incinerator
(Ghianti, S.). The use of this unit to burn wastes containing dioxin seems
much more likely than the use of the two G.E. incinerators, since
transportation of the waste to the Occidental Facility would be minimal.

With regard to ocean incineration, there are currently no existing
designated or approved burn sites in U.S. coastal waters. Therefore, the
status of this technology is uncertain at best. Chemical Waste Management
(CWM) has conducted research burns and trial burns for PCB-containing wastes
and dioxin-containing waste. However, there has been strong public opposition
to full-scale operation (Bond, 1984; HMIR, 1985a and b; HMIR, 1986a and b).
CWM has applied for a permit to conduct another research burn of PCBs at sea
to address public concerns about emissions and toxic effects that have been
raised by the Science Advisory Board in a 1985 report (Brown, 1986; HMIR,
1985a). The proposed test plan includes obtaining more data on DREs for PCBs
(99.9999 percent-DRE is required by permit) (Brown, 1986). The Science
Advisory Board believes that prior ocean incineration testing has not
conclusively characterized emissions because effluent streams were only
analyzed for a limited range of constituents. During the proposed burn, a
full GC/MS scan will be performed to address this concern (Brown, 1986).

Additionally, the Science Advisory Board suggested that synergistic
and/or antagonistic reactions between emitted compounds may increase the
toxicity of emissions beyond that which would be expected from the toxicities
attributed to each compound individually. Therefore, during the proposed
burn, concentrated samples collected from the stack effluent will be injected
into biological organisms (in a laboratory environment) to determine the
concentrations at which the effluent emissions will cause toxic effects
(Brown, 1986).

The U.S. EPA published the proposed permit regulations on December 16,
1985. The comment period ended on February 15, 1986. A final decision is
expected in May 1986.

4.4   FLUIDIZED BED/CIRCULATING FLUIDIZED BED (CFB) SYSTEM

4.4.1   Process Description

The fluidized bed incinerator uses high temperature oxidation under controlled conditions to destroy organic constituents in liquid, gaseous, and solid waste streams. It is typically used for slurries and sludges.

As shown in Figure 4.4.1, a typical fluidized bed incinerator consists of a vertical refractory-lined cylindrical vessel containing a bed of inert granular material (typically, sand) on a perforated metal plate. The waste (in the form of either gas, liquid, slurry, or sludge) is usually injected into or just above the stationary bed. The granular bed particles are fluidized by blowing air upward through the medium. The resulting agitation ensures intimate mixing of all waste material with combustion air (McGaughey, et al., 1984; Bonner, 1981).

A burner located above the bed is used to heat the bed to start-up temperature. The large mass and high heat content of the bed causes the waste to rapidly combust which, in turn, transfers heat back to the bed. The maximum temperature of the granular bed is limited by the softening point of the bed material (for sand this temperature is 1100°F). The residence time of waste material in the bed typically ranges from 12 to 14 seconds for liquid wastes. The solid uncombustible materials in the waste become finely suspended particulate matter which is separated in a cyclone while the exhaust gases pass through an afterburner to destroy vapor-phase residuals (McGaughey, et al., 1984; Bonner, 1981).

Waste Tech Services, Inc. has developed a Low-Temperature Fluidized Bed that functions similarly to the conventional fluidized bed except that a higher air volume is forced through the bed material (Rasmussen, 1986; Freeman, 1985). Also, the bed is composed of a mixture of a granular combustion catalyst and limestone. Limestone is continuously added to the bed and the bed material is periodically drained from the vessel. A multicyclone system employing a baghouse to clean the flue gas is used for air-pollution control. The Waste-Tech fluidized bed is able to operate at lower temperatures than conventional fluidized beds and also has reduced supplemental fuel requirements (Rasmussen, 1986; Freeman, 1985).

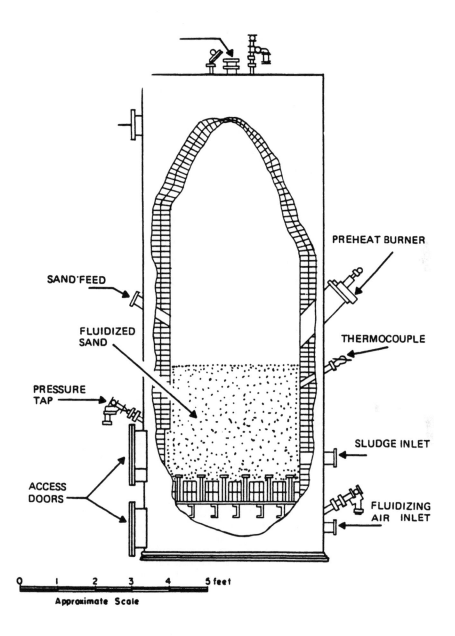

Figure 4.4.1.   Cross-section of Fluidized-Bed Furnace [U.S. EPA, 1979].

Another modification of the conventional fluidized bed technique that has been developed is the Circulating Fluidized Bed Combuster (Figure 4.4.2). It utilizes contaminated soil as the bed material and air flow rates 3 to 5 times greater than conventional systems (Rickman, et al., 1985; Vrable, et al., 1985a and b; Barner, 1985). The high air flow causes increased turbulence which allows for efficient combustion at much lower operating temperatures without requiring the use of an afterburner. A comparison of the circulating fluidized bed combustor with the conventional fluidized bed is shown in Table 4.4.1.

The startup combustor burner consists of a natural gas fuel system (Rickman, et al., 1985; Vrable, et al., 1985a and b; Barner, 1985). It has a 4 to 6 hour cold startup period, and an approximately 30 minute hot restart (with a refractory temperature at or greater than 1400°F). The startup burner is generally idle during waste burning unless the waste feed is interrupted and it is required to maintain a low combustor temperature. The combustor is a carbon-steel tube with refractory lining which consists of an erosion-resistant inner layer and a thermal insulating outer layer. Prior to being injected into the combustion chamber the waste feed is mixed with hot recirculating solids from the cyclone. Both the waste feed and the recirculated solids are introduced into the combustion chamber. Liquid and slurry waters are pumped from stirred tanks whereas a metering screw is used to convey solids and sludges.

The combustor has primary and secondary air ports through which fluidizing air is provided by a constant-speed, motor-driven forced-draft fan (GA Technologies, 1985; Rickman, et al., 1985). The high air velocity (15 to 20 feet/second) entrains both the bed and the combustible waste which rise through the reaction zone to the top of the combustion chamber and pass into a hot cyclone.

The cyclone is constructed of carbon-steel and lined with castable refractory lining. The function of the cyclone is to separate bed material from the combustion gases and recirculate these solids to the combustion chamber. The hot combustion gases flow to an off-gas heat exchanger where they are cooled to 375°F and then directed to baghouse filters to remove any residual products of incomplete combustion (GA Technologies, 1985; Rickman, et al., 1985).

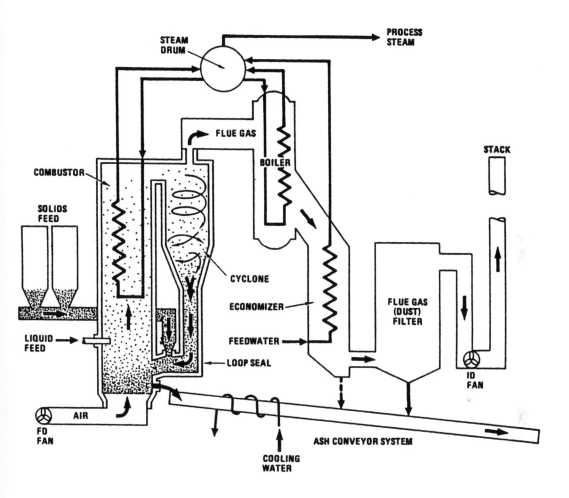

Figure 4.4.2.   Schematic of Circulating Bed Combustor.
Source: GA Technologies, 1985.

TABLE 4.4.1.   COMPARISON OF CONVENTIONAL FLUIDIZED BED WITH
CIRCULATING FLUIDIZED BED COMBUSTOR

Condition	Circulating fluidized bed	Conventional fluidized bed
Feeding		
No. of Inlets	1-solid; 1-liquid	5-solid; 5-liquid
Sludge Feeding	Direct	Filter/Atomizer
Solids Feed-size	<1 in.	<0.5-0.25 in.
Pollution Control		
POHCs	In moderate temp. combustor	In high temp. combustor or afterburner
Cl,S,P	Dry limestone in combustor	Downstream scrubber
Upset Response	Slump bed; no release	Bypass scrubber pollution release
Effluent	Dry Ash	Wet Ash Sludge
Efficiency		
Thermal	>78%	>75%
Carbon	>98%	>90%

Reference: Rickman, et al., 1985

Bottom ash is continuously removed from the combustor, cooled to approximately 200°F using a water-cooled screw conveyor and solidified or drummed.

Auxiliary fuel may also be injected into the bed if required to maintain temperature. Retractable bayonet heat exchangers are used when combustion heat must be removed from the combustor. These heat exchangers are constructed of stainless steel and use recirculated water as a coolant. Adjustment of the heat exchanger area enables control of the combustor temperatures (GA Technologies, 1985; Rickman, et al., 1985).

Restrictive Waste Characteristics--

The fluidized bed incineration technique is not well-suited for irregular, bulky solids, tarry solids, or wastes with a high fusible ash content (Freeman, 1985; GCA, 1985). Formation of eutectics (compounds with low melting or fusion temperatures) can result in bed fouling. Problems caused by wastes with low ash fusion temperatures can be avoided by keeping operating temperatures below the ash fusion level or by using chemical additives to raise the ash fusion temperature. Waste containing bulky or irregular solids may require pretreatment in the form of drying, shredding, and sorting prior to entering the reactor.

Labor utilization is high since regular preparation and maintenance of the fluid bed must be performed. These costs can increase dramatically if it becomes difficult to remove residual materials from the bed.

Operating Parameters--

The diameter of the fluidized bed unit typically ranges from a few meters to 15 meters. Operating temperatures normally range from 450°C to 980°C and are limited by the softening point of the bed media (1100°C for sand) (McGaughey, et al., 1984; GCA, 1985). Residence times are generally on the order of 12 to 14 seconds for a liquid hazardous waste.

4.4.2   Technology Performance Evaluation

Fluidized beds have been used to treat municipal wastewater treatment plant sludges, oil refinery waste, pulp and paper mill waste, pharmaceutical wastes, phenolic wastes, and methyl methacrylate.   Pilot-scale demonstrations have been performed for other hazardous wastes.   Currently, there are more than 25 circulating bed combustors operating in the U.S. and Europe.   However, there are currently no units operating commercially as hazardous waste incinerators (Freeman, 1985; Rickman, et al., 1985).

The low-temperature fluidized bed combustor (designed by Waste Tech Services, Incorporated) was used to conduct trial burns on soil contaminated with carbon tetrachloride and dichloroethane (Freeman, 1985).   Only four nines DRE was demonstrated.   The results of these tests are summarized in Table 4.4.2.

GA Technologies has conducted trial burns on its pilot scale circulating bed combustor using chlorinated organic liquid wastes.   The combustor was operated at 1540 to 1600°F with a gas velocity of 11 to 12 feet/second and 45 to 60 percent excess air.   Limestone was injected into the incinerator with the liquid waste feed to prevent the formation of HCl by capturing the chlorides formed.   The following results were obtained (Rickman, et al., 1985; Chang and Sorbo, 1985):

NO_x emissions	40 ppm (average)
SO_2 emissions	250 to 350 ppm
CO emissions	1000 ppm
Chloride Capture	99%

Flue Gas Emissions (%-DRE)
- Ethylbenzene   >99.99
- 1,1,2-trichloroethane   >99.99
- 1,2-dichloroethane   >99.99
- 1,1-dichloroethylene   >99.99
- 1,2-transdichloroethylene   >99.99
- vinyl chloride   >99.99
- toluene   >99.99
- benzene   >99.99

TABLE 4.4.2. RESULTS OF LOW-TEMPERATURE FLUIDIZED BED TRIAL BURN USING
              SOIL CONTAMINATED WITH CARBON TETRACHLORIDE
              AND DICHLOROETHANE

Condition	Run 1	Run 2
Fluidized Bed Temperature	850°C	850°C
Combustion Vessel Exit Temperature	650°C	650°C
Vessel Residence Time	1.3 sec	1.3 sec
Feed Rate:		
Soil	10.5 kg/h	6.17 kg/h
Carbon Tetrachloride	0.32 kg/h	0.40 kg/h
Dichloroethane	0.035 kg/h	0.044 kg/h
Destruction Efficiency:		
Carbon Tetrachloride	99.998%	99.996%
Dichloroethane	99.998%	99.997%

Reference: Freeman, 1985; Rasmussen, 1986.

The pilot-scale unit was also used to conduct trial burns on PCB-contaminated soil (Rickman, et al., 1985; Chang and Sorbo, 1985). An auxilliary fuel was used to maintain bed temperature at 1600 to 1800°F. A destruction efficiency exceeding six nines (99.9999 percent) was achieved. A summary of the test conditions and results is given in Table 4.4.3.

4.4.3. Costs of Treatment

The costs for the conventional fluidized bed are dependent on fuel requirements, scale, and site conditions. However, the costs are generally comparable with conventional rotary kiln incineration technology. Waste-Tech Services, Inc. lists the costs in Table 4.4.4 as being typical for their low-temperature fluidized bed (Freeman, 1985).

Costs for the circulating bed combustor vary according to the size of the incineration unit, and the type of waste being processed. Estimated costs for a 25 million Btu/hr unit are given in the Table 4.4.5 (Freeman, 1985):

4.4.4 Process Status

Currently, there are several fluidized bed combustors operating worldwide. Although fluidized beds have been used in various industries, at the present time there are not any fluidized beds operating commercially as hazardous waste incinerators. However, the fluidized bed, particularly the circulating fluidized bed, appear to have significant potential for future use in the destruction of hazardous wastes.

The low-temperature fluidized bed developed by Waste Tech Services requires additional testing and/or development, but could potentially be used for the destruction of dioxin contaminated wastes.

A stationary pilot scale circulating fluidized bed unit capable of incinerating a ton per hour of hazardous waste is in operation at the GA Technologies test facility. A transportable incinerator has also been constructed for use in onsite demonstrations on PCBs contaminated wastes. Although dioxin trial burns have not been conducted, GA Technologies would be interested in performing dioxin testing if funding were available (Jensen, 1986).

TABLE 4.4.3.   PCB-CONTAMINATED SOIL TRIAL BURN TEST CONDITIONS
AND RESULTS FOR CIRCULATING BED COMBUSTOR

Condition/result	Trial 1	Trial 2	Trial 3
PCB Concentration in soil (ppm)	11,000	12,000	9,800
Soil Feed Rate (lb/hr)	325	410	325
Combustion Temperature (°F)	1800	1800	1800
Surface Velocity in Combustion Chamber (ft/sec)	18.7	18.7	18.1
% Excess Oxygen	7.9	6.8	6.8
%-DRE	>99.9999	>99.9999	>99.9999
PCB in bed ash (ppm)	0.0035	0.033	0.186
PCB in fly ash (ppm)	0.066	0.010	0.032
Dioxin/Furan in ash (ppm)	ND	ND	ND
% Combustion Efficiency	99.94	99.95	99.97
NO_x (ppm)	26	25	76
CO (ppm)	35	28	22
HCl (ppm)	57	202[a]	255[a]
Particulates (g/dry cu. ft)	0.09[b]	0.04	0.002

[a]High values resulted from intermittent limestone addition.

[b]Obtained from makeup test for particulates only.

Reference: Rickman, et al., 1985; Chang and Sorbo, 1985.

TABLE 4.4.4.   WASTE-TECH FLUIDIZED BED COSTS

Item	Cost
Operating Labor	0.0084 $/lb
Consumables and Utilities	0.0138 $/lb
Nonlabor (capital depreciation, siting cost, maintenance mat'ls, insurance, tax overhead)	0.0116 $/lb
Limestone for Chlorine Removal, Waste Excavation, Ash Disposal, etc.	0.043 $/lb
TOTAL COST	$150/ton

Note: These cost estimates are for a 50 sq.ft. system with a throughput of 9,200 lb/hr for soils having 2 percent organics and 5 percent moisture content.

TABLE 4.4.5.   CIRCULATING FLUIDIZED BED COSTS

Feed Type	Installed Capital Costs	Annual Operating Costs	Total Cost per Unit of Feed
Chlorinated Organic Sludge	$2.0 million	$0.25 million	$60/ton
Contaminated Soil	$1.8 million	$0.35 million	$27/ton
Wet Sludge	$1.8 million	$0.35 million	$32/ton

Note: Costs are based on the use of a 25 million Btu/hr unit.

4.5 HIGH TEMPERATURE FLUID WALL (HTFW) DESTRUCTION - ADVANCED ELECTRIC REACTOR

4.5.1 Process Description [Lee, Schofield, and Lewis, 1984; Schofield, Scott and Dekany, 1985; Weston, Inc. 1985.]

The HTFW reactor was originally developed by Thagard Research of Costa Mesa, California. The J.M. Huber Corp. of Borger, Texas has developed proprietary modifications to this original design. This reactor, called the Advanced Electric Reactor (AER), is shown in Figure 4.5.1. The reactor is a thermal destruction device which employs radiant energy provided by electrically heated carbon electrodes to heat a porous reactor core. The core then radiates heat to the waste materials. The reactor core is isolated from the waste by a blanket of gas formed by nitrogen flowing radially through the porous core walls.

The only feed streams to the reactor are the waste material and the inert nitrogen gas blanket. Therefore, the destruction is by pyrolysis rather than oxidation. Because of the low gas flow rate and the absence of oxygen, long gas phase residence times can be employed, and intensive downstream cleanup of off gases can be achieved economically.

Destruction via pyrolysis instead of oxidation significantly reduces the concentrations of typical incineration products such as carbon monoxide, carbon dioxide, and oxides of nitrogen. The principal products formed during treatment of soil contaminated with TCDD are hydrogen, chlorine (if calcium oxide is added to the reactor, calcium chloride is formed instead), hydrochloric acid, elemental carbon, and free-flowing granular material (Schofield, et. al., 1985; Boyd, et al., 1986; GCA, 1985)

A process flow diagram for the AER is shown in Figure 4.5.2. The waste, if it is a solid, is released from an air tight feed bin through a metered screw feeder into the top of the reactor. If it is a liquid, it is fed by an atomizing nozzle into the top of the reactor. The waste then passes through the reactor where pyrolysis occurs at temperatures of approximately 4500°F in the presence of nitrogen gas. Downstream of the reactor, the product gas and waste solids pass through two postreactor treatment zones, the first of which is an insulated vessel which provides additional high temperature (2000°F) and

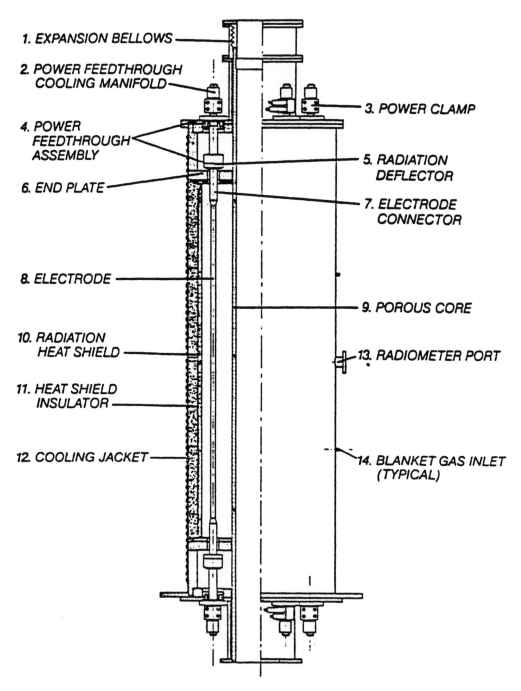

1. EXPANSION BELLOWS

2. POWER FEEDTHROUGH COOLING MANIFOLD

3. POWER CLAMP

4. POWER FEEDTHROUGH ASSEMBLY

5. RADIATION DEFLECTOR

6. END PLATE

7. ELECTRODE CONNECTOR

8. ELECTRODE

9. POROUS CORE

10. RADIATION HEAT SHIELD

13. RADIOMETER PORT

11. HEAT SHIELD INSULATOR

12. COOLING JACKET

14. BLANKET GAS INLET (TYPICAL)

Figure 4.5.1.   Advanced Electric Reactor [Huber].

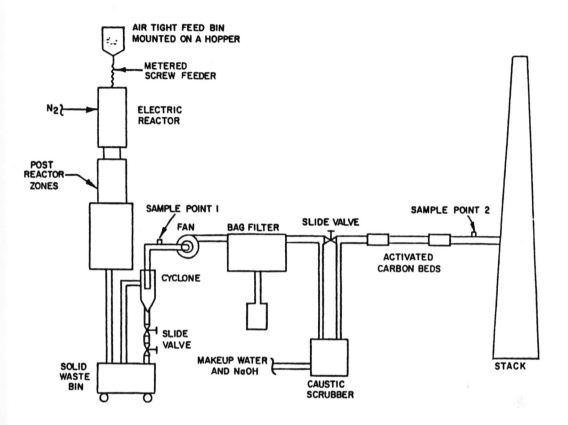

Figure 4.5.2.   High temperature fluid wall process configuration for the
destruction of carbon tetrachloride [Huber].

residence time (5 seconds). The second postreactor treatment zone is water-cooled, and its primary purpose is to cool the gas prior to downstream particulate cleanup.

Off gas cleaning equipment includes a cyclone to collect particles which do not fall into the solids bin, a bag filter to remove fines, an aqueous caustic scrubber for acid gas and free chlorine removal, and two banks of five parallel activated carbon beds in series for removal of trace residual organics and chlorine.

The stationary pilot scale reactor which has been used for testing various wastes at their Borger, Texas facility consists of a porous graphite tube, 1 foot in diameter and 12 feet high, enclosed in a hollow cylinder with a double wall cooling jacket. This pilot unit is capable of processing 5000 tons/yr of waste. Huber also has a 3 inch diameter mobile unit which has been transported to hazardous waste sites for testing purposes. Test results are described below.

Restrictive Waste Characteristics--

The AER cannot currently handle two-phase materials (i.e., sludge); it can only burn single-phase materials consisting of solids, or liquids, or gases alone (Schofield, 1985; Boyd, 1986). Generally, a solid feed must be free flowing, nonagglomerating, and smaller than 100 mesh (less than 149 micrometers or 0.0059 inches) (GCA, 1985; Shofield, 1985). However, depending on the required destruction, solids larger than 100 mesh (but smaller than 10 mesh) may be suitable. Soils should be dryed and sized before being fed into the reactor.

Also, the Huber process is not cost competitive with standard thermal destruction techniques (such as the rotary kiln) for materials with a high Btu content (Schofield, 1985; Boyd, 1986). It is cost-effective for wastes with a low Btu content (i.e., PCBs and dioxin) because unlike standard thermal destruction techniques, the Huber process does not require supplementary fuels to obtain the necessary Btu content for incineration.

Operating Parameters--

Typical operating parameters for the Advanced Electric Reactor are summarized below (Freeman, 1985; Schofield, et al., 1985; Boyd, et al.,1986):

Residence Time (100 mesh solids)	0.1 seconds
Gas Flow Rate	500 scfm for 150 ton day scale
Gas phase Residence Time (at 2500° F)	5 seconds

4.5.2 Technology Performance Evaluation

In 1983, Thagard conducted a series of tests on PCB-contaminated soils using a 3-inch diameter research reactor (Horning and Masters, 1984; Freeman, 1985). The results of these tests showed an average DRE of 99.9997 percent. The destruction efficiency was found to be independent of the feed rate in the 50 to 100 g/min range at 2343°C. Pyrolysis products other than carbon and hydrogen chloride were not detected using a GC with electron capture detection. It was concluded that the method for dispersing the feed into the reactor needed improvement.   Problems with slagging in the reactor occurred that were believed to be related to the small diameter of the reactor and also to the design of the fluid wall flow. After modifications, additional tests on a 6-inch prototype reactor were conducted by Thagard using hexachlorobenzene dispersed on carbon particles; 99.99991 percent destruction efficiency was achieved (Horning and Masters, 1984).

J.M. Huber Corporation purchased the patent rights and made further improvements to the process (Boyd, 1986). The J.M. Huber Corporation then began tests in its stationary reactor system which has a diameter of 12 inches. Included in this system are: an insulated post-reactor vessel, a water-jacketed cooling vessel, a cyclone, a baghouse, a wet scrubber, and an activated carbon bed (Boyd, et al., 1986; Schofield, et al., 1985; Freeman, 1985). Several research burns have been conducted with this system (Schofield, et al., 1985). Results and operating parameters for pertinent burns are summarized in Table 4.5.1.

TABLE 4.5.1.   SUMMARY OF OPERATING PARAMETERS AND RESULTS
FOR HUBER AER RESEARCH/TRIAL BURNS

Condition	PCBs (Sept. 1983)	CC14 (May 1984)	Dioxins (Oct/Nov 1984)
Reactor Core Temperature (F)	4100	3746–4418	3500–4000
Waste Feed Rate (lb/min)	15.5–15.8	1.1–40.8	0.4–0.6
Nitrogen Feed Rate (scfm)	147.2	104.3–190.0	6–10
%-DRE	>99.99999	>99.9999	>99.999

Reference: Schofield, 1984; Roy F. Weston, 1985.

A series of four trial PCB-burns were conducted during September 1983
using a synthesized mixture of Aroclor 1260 and locally available sand to
obtain a total concentration of 3000 ppm PCBs (Schofield, et al., 1985;
Freeman, 1985). After treatment, the sand had a PCB content ranging from
0.0001 to 0.0005 ppm (0.1 to 0.5 ppb). The destruction and removal efficiency
was measured to be 99.99960 to 99.99995 percent. Additional studies were
conducted with the 12 inch diameter reactor using soils contaminated with
octachlorodibenzo-p-dioxin (OCDD) and carbon tetrachloride. Seven nines DRE
(99.99999 percent) were reportedly achieved at feed rates up to 2500 lbs/hr.
J.M. Huber also maintains a 3 inch diameter mobile reactor which was used for
trial burns on 2,3,7,8-TCDD-contaminated soil at Times Beach, Missouri in
November 1984 (Roy F. Weston, 1985). The 2,3,7,8-TCDD levels in the soils
ranged from 67.9 to 99.8 ppb. A total of 63.58 lbs of soil were processed at
temperatures ranging from 2260° to 2315°C.

Greater than five nines (99.999 percent) destruction and removal
efficiency (DRE) was demonstrated during the Times Beach trial burns (Roy F.
Weston, 1985). Higher DREs could not be demonstrated due to the inability of
the instrument detection limits to compensate for the relatively low quantity
of contaminated soil (∼79 ppb in 63.58 lbs or 0.002 grams). 2,3,7,8-TCDD
concentrations were below detection limits in the treated soil (<0.11 ppb),
in the baghouse catch (<0.55 ppb), and in the stack emissions (<0.71 ppb).
Chlorinated organics were not detected in the stack emissions (at a detection
limit of 25 ppb), and gaseous emissions of particulates were within EPA
standards.

During the Times Beach trial burns, difficulties were encountered in the
soil preparation (such as maintaining dryness and properly sized soils) which
caused Huber Corporation to delay the construction of a 50,000 ton/yr reactor
(Technical Resources, Inc., 1985). The U.S. Air Force recently conducted
studies related to the pretreatment of dioxin contaminated soils using the
3 inch mobile AER system at Gulfport, Mississippi (Boyd, 1986). Although the
testing was completed in June 1985, the results are currently being reviewed
and a final report should be available in the summer of 1986. The studies
consisted primarily of soil pretreatment techniques for more efficient
operation of the reactor (Boyd, 1986).

4.5.3   Costs of Treatment

Operating costs will vary depending on the quantity of material to be processed and the characteristics of the waste feed (Lee, et al., 1984). Pretreatment may be necessary for bulky wastes having a high moisture content. Typical energy requirements for normal soil range from 800 to 1000 kwh/ton.

Cost estimates for processing a site containing more than 100,000 tons of waste material were approximately $365 to $565/ton in 1985 (Lee, et al., 1984; Freeman and Olexsey, 1986). The cost breakdown for this estimate was 12 percent for maintenance, 7 percent labor, 29 percent energy, 18 percent depreciation and 34 percent for other costs (permitting, setup, post-treatment, etc.). These costs have recently been updated. The new costs are expected to be released in May 1986 (Boyd, 1986).

4.5.4   Process Status

The J.M. Huber Corporation purchased the patent rights from Thagard Research Corporation. Huber then modified the design of the reactor (primarily the feed tube and the core design) to improve the efficiency of the reactor, extend the lifetime of the electrodes and core material, and to reduce sticking of vitreous material on the core walls (which lowers efficiency) (Boyd, et al., 1986).

Huber maintains two fully-equipped reactors at their pilot facility in Borger, Texas (Schofield, et al., 1985). The smaller reactor, which is equipped for mobile operation, has a 3-inch core diameter and a capacity of 0.5 lb/min.. The larger reactor is commercial scale with a 12-inch core diameter and a capacity of 50 lb/min. Both of these reactors are used primarily for research purposes. In May 1984, the Huber reactor was certified by the EPA under TSCA to burn PCB wastes. Recently, the U.S. EPA and the Texas Water Commission jointly issued J.M. Huber Corporation a RCRA permit which authorizes the incineration of any non-nuclear RCRA hazardous waste (including dioxin-containing wastes) in the Huber Advanced Electric Reactor (AER)(HMIR, 1986). This was the first commercial permit issued under

RCRA for treating dioxin-containing wastes. The J.M. Huber Corporation intends to use the permit for research and development of a full-scale transportable AER.

Huber does not intend to operate a hazardous waste disposal operation, but rather to construct and market stationary and/or mobile units for use by companies or organizations involved in hazardous waste destruction (Boyd, 1986).

4.6 INFRARED DESTRUCTION (Shirco)

4.6.1 Process Description (Daily, 1986; Shirco, 1985; Freeman and Olexsey, 1986; HMIR, 1986; Technical Resources Inc., 1985; Daily, 1985)

Shirco Infrared Systems, Inc. has developed a portable infrared incineration system, which can be transported in a 45 ft trailer. The major components of the system include a feed metering system, an infrared primary chamber furnace, a combination propane-fired/infrared secondary chamber, a venturi scrubber system, blower and heating control systems, and a monitoring and control system.

The waste material is fed by bucket or inclined conveyor onto a metering conveyor which controls the amount and rate of waste feed into the primary furnace. The primary furnace chamber is constructed of carbon steel, lined with multiple layers of ceramic fiber blanket-insulation mounted on stainless steel studs and retained with ceramic fasteners. The external dimensions of the primary chamber are 2.5 ft x 9 ft x 7 ft, and it weighs (installed) 3,000 lbs. Infrared heating elements, consisting of silicon carbide rods with external electrical connections at each end, are spaced along the length of the furnace. The chamber can be heated to temperatures ranging from 500 to 1,050°C. Residence times for the feed material are variable ranging between 10 and 180 minutes. The temperatures and times will depend on the characteristics of the waste.

Following combustion, the ash (or processed material) is conveyed to the end of the furnace where it drops off the belt and passes through a chute into an enclosed, tapered hopper. A discharge screw conveyor controls transport of the discharged material from the hopper into sealed collection drums.

Combustion air is forced through a combustion air preheater and then injected at 10 points along the length of the primary chamber furnace. Depending on the waste characteristics, the exhaust gases may be directed to a secondary combustion chamber to complete gas-phase combustion reactions.

The secondary chamber is a rectangular carbon steel box lined with a ceramic fiber blanket insulation. The secondary chamber weighs 1,500 lbs and has external dimensions of 3 ft x 9 ft x 3 ft. Combustibles in the gas are ignited via a propane-fired burner and are maintained at a predetermined setpoint temperature using an array of silicon carbide heating elements which also are arranged to enhance gas turbulence. Turbulence is also provided by combustion air from the blower which is injected into the secondary chamber through two offset jets on each side of the chamber. Combustion residence times typically range from 1.5 to 2.2 seconds, with a process temperature capability of up to 2,300°F.

Exhaust gases from the secondary chamber pass through a wet scrubber for removal of particulates. The scrubber also cools the gases (from incoming temperatures of 1,000 to 2,300°F) to saturation temperature (generally 180°F). The gases can be subcooled to lower temperatures, but this requires a significantly greater volume of water. A pump is used to recirculate the scrubber water from a sump tank to the scrubber. Lime may be added to the sump tank to remove any acid materials formed. A blower is used to direct the exhaust gases through a 10 ft exhaust stack.

Available sampling points include two standard sampling ports in the exhaust stack, and sampling ports in the primary and secondary chamber exhaust ducts. In addition, temperatures, pressures, and flow rates are measured and shown on a master control panel.

4.6.2 Technology Performance Evaluation

In early June 1985, the Shirco portable pilot test unit was taken to Times Beach, Missouri to conduct trial burns on 2,3,7,8-TCDD-contaminated soils (ERT, 1985: Daily, 1986). The system was set up and ready for operations within a few hours of arrival at the site. The testing was conducted over a 2-day period followed by equipment decontamination activities. Two tests were performed, each with differing operating parameters. The operating parameters and the results of each test are shown in Table 4.6.1.

Emissions samples were collected over a 7-hour period for Test 1 (30-minute residence time), and over a 2.5-hour period for Test 2 (15-minute residence time)(ERT, 1985; Daily, 1986). Continuous samples of the thermally treated soil were also collected during each run. Dioxin was not detected (based on analytical detection limits) in the treated residual material for either test run. Sufficient gas sample was collected to demonstrate destruction and removal efficiencies (DREs) for both runs that

TABLE 4.6.1   OPERATING PARAMETERS AND RESULTS FOR SHIRCO INFRARED
DESTRUCTION PILOT TESTS

Condition	Test 1	Test2
TCDD in Feed (ng/g)	227	156
Solid Phase Residence Time (min)	30	15
Solid Feed Rate (lb/hr)	47.68	48.12
Primary Chamber Temp.-Zone A (°F)	1560	1490
Primary Chamber Temp.-Zone B (°F)	1550	1490
Secondary Chamber Temperature (°F)	2250	2235
Emissions Sampling Duration (hours)	7	2.5
Particulate at 7% O_2 (gr/dscf)	0.0010	0.0002
Gas Phase DRE of 2,3,7,8,-TCDD	>99.999996	>99.999989
at Detection Limit (picograms)	14	8.4
Ash Analysis for 2,3,7,8-TCDD	ND	ND
at Detection Limit (ppt)	38	33

Reference: ERT, 1985; Daily, 1986.

exceeded the required 99.9999 percent DRE (when calculated at the detection limits). Particulate emissions were well below the 0.08 gr/dscf EPA regulation requirement.

4.6.3 Costs of Treatment

Preliminary estimates of the operating costs for infrared incineration have been stated to be under $200/ton (Daily, 1985). This estimate includes the cost of insurance, obtaining permits, labor, and energy requirements. Costs for excavation and disposal are not included in this estimate. Construction costs for the transportable incinerator range from 2 to 3 million dollars.

4.6.4 Process Status

The results obtained at The Times Beach Trial Burn have demonstrated that the Shirco Infrared Process is a viable technology for dioxin decontamination. The Shirco Infrared Incinerator is expected to be commercially available in the near future (Johanson, 1986; Hill, 1986).

4.7 PLASMA ARC PYROLYSIS

4.7.1 Process Description

Operation and Theory--

In this process waste molecules are destroyed by the action of a thermal plasma field. The field is generated by passing an electrical charge through a low pressure air stream, thereby ionizing the gas molecules and generating temperatures up to 10,000°C.

A flow diagram of the plasma pyrolysis system is shown in Figure 4.7.1. The plasma device is horizontally mounted in a refractory-lined pyrolysis chamber with a length of approximately 2 meters and a diameter of 1 meter. Liquid wastes are injected through the colinear electrodes of the plasma device where the waste molecules dissociate into their atomic elements. These elements then enter the pyrolysis chamber which serves as a mixing zone where the atoms recombine to form hydrogen, carbon monoxide, hydrogen chloride and particulate carbon. The approximate residence times in the atomization zone and the recombination zone are 500 microseconds and 1 second, respectively. The temperature in the recombination zone is normally maintained at 900-1,200°C (Barton, 1984).

After the pyrolysis chamber, the product gases are scrubbed with water and caustic soda to remove hydrochloric acid and particulate matter. The remaining gases, a high percentage of which are combustible, are drawn by an induction fan to the flare stack where they are electrically ignited. In the event of a power failure, the product gases are vectored through an activated carbon filter to remove any undestroyed toxic material.

The treatment system that is currently being used for testing purposes is rated at 4 kg/minute of waste feed or approximately 55 gal/hour. The product gas production rates are 5-6 m^3/minute prior to flaring. To facilitate testing, a flare containment chamber and 30 ft stack have also been added to the system. The gas flow rate at the stack exit is approximately 36 m^3/minute (Kolak, Barton, Lee, Peduto, 1986).

A major advantage of this system is that it can be moved from waste site to waste site as desired. The entire treatment system, including a laboratory, process control and monitoring equipment, and transformer and switching equipment, are contained on a 45 ft tractor-trailer bed (Barton, 1984).

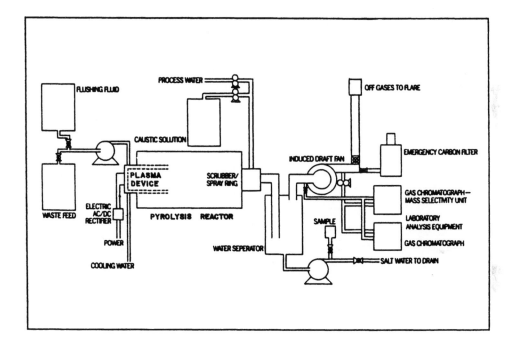

Figure 4.7.1.   Pyroplasma process flow diagram.

Source: Kolak, et al., 1986.

Two residual streams are generated by this process. These are the
exhaust gases released up the stack as a flare, and the scrubber water
stream. Since the product gas (after scrubbing) is mainly hydrogen, carbon
monoxide, and nitrogen, it burns with a clean flame after being ignited.
Analysis of the flare exhaust gases, presented in the following section,
indicates virtually complete destruction of toxic constituents.

The scrubber water stream is composed mainly of salt water from
neutralization of HCl and particulates, primarily carbon. Analyses of the
scrubber water for the waste constituent of concern (e.g., carbon
tetrachloride (CCl_4) and PCB in the feed material) have shown that the
constituents were present at low ppb concentrations. The quality of scrubber
water generated would depend on the water feed rate and corresponding product
gas and scrubber waste flowrates. During a test in which 2.5 kg/min of waste
containing 35 to 40 percent CCl_4 was fed to the reactor, a scrubber water
effluent flowrate of 30 l/minute was generated (Kolak, Barton, Lee,
Peduto, 1986).

Restrictive Waste Characteristics--
The reactor as it is currently designed can only be used to treat liquid
waste streams with viscosities up to that of 30 to 40 weight motor oils.
Particulates are removed by a 200 mesh screen prior to being fed into the
reactor. Contaminated soils and viscous sludges cannot be treated. The TCDD
wastes for which this technology has potential include nonaqueous phase
leachate such as that which has been generated at the Love Canal and Hyde Park
Landfills, unused liquid herbicide solutions such as herbicide orange, and
possibly still bottoms from herbicide production.

4.7.2   Technology Performance Evaluation

The plasma arc system has been tested using several liquid feed materials
including carbon tetrachloride (CCl_4), polychlorinated biphenyls (PCBs), and
methyl ethyl ketone (MEK). It has not been tested on wastes or other materials
contaminated with TCDD. However, because of the structural similarity between
TCDD and PCBs, the data presented should provide some indication as to the
potential of this technology towards destroying TCDD.

Table 4.7.1 presents the results of three test burns conducted in Kingston, Ontario using carbon tetrachloride in the feed material. The carbon tetrachloride was fed to the reactor along with ethanol, methyl ethyl ketone, and water at a rate of 1 kg of CCl_4/minute. The duration of each of these tests was 60 minutes, and stack gas flowrates and temperatures averaged 32.5 dry standard cubic meter/minute (dscm/min) and 793°C, respectively. As can be seen in the table, the destruction and removal efficiency (DRE) of CCl_4 in each of the tests exceeded six nines which is the required DRE when incinerating wastes containing TCDD. In addition, the concentration of HCl in exhaust gases was less than the upper limit of 1.8 kg/hr required by RCRA guidelines. The only possible area of concern is that the concentration of CCl_4 in the scrubber water is greater than 1 ppb. Proposed regulations for treatment of TCDD require that the residuals have less than 1 ppb of TCDD for the residual to be nonhazardous (Kolak, Barton, Lee, Peduto, 1986).

Table 4.7.2 contains results of other tests conducted using PCB in the feed material. During startup of the unit, a mixture of MEK and methanol (MeOH) was fed to the reactor. Once the exhaust gas attained a temperature of 1,100°C, the waste feed was switched to a blend of PCB, MEK, and MeOH. Typical operating parameters for these tests are presented in Table 4.7.2. Stack gases and scrubber effluents were analyzed for dioxins, furans and benzo(a)pyrene, in addition to PCBs. As indicated in Table 4.7.3, the concentrations of these constituents in each of the residual streams is extremely low. The concentrations of dioxins in the scrubber water and the stack gases are both in the low parts per trillion range. As far as PCBs are concerned, the destruction and removal efficiency in each of the tests was greater than 6 nines, and in some cases reached 8 nines.

4.7.3 Costs of Treatment

The approximate capital cost of a unit similar to the one tested would be in the range of 1 to 1.5 million dollars (Plottner, 1986). More accurate figures will be available once a commercial unit has been built. Nonetheless, some general operating costs associated with using the reactor can be estimated. One of the major costs would be the electrical power required to generate the plasma torch. If, for example, the currently built unit were used to destroy nonaqueous leachate collected at the Hyde Park and Love Canal

TABLE 4.7.1.   CARBON TETRACHLORIDE TEST RESULTS

Parameter	Test 1	Test 2	Test 3
Chlorine Mass Loading (%)	35	40	35
Scrubber Effluent			
CCl_4 (ppb)	1.27	5.47	3.26
mg/hr	2.29	9.85	5.87
Flare Exhaust			
CCl_4 (ppb)	0.83	0.43	0.63
mg/hr	12.1	4.9	7.2
NO_x			
ppm(v/v)	106	92	81
lbs/hr	1.02	0.69	0.02
CO			
ppm(v/v)	48	57	81
lbs/hr	0.28	0.26	0.37
HCl			
mg/dscm *	(1)	137.7	247.7
kg/hr	(1)	0.25	0.44
Destruction Removal Efficiency	99.99998	99.99998	99.99998

(1)   sample taken was invalidated due to plugging of sampling apparatus

*mg/dscm = milligrams per dry standard cubic meter.

TABLE 4.7.2. TYPICAL OPERATING DATA FOR PCB TESTS (ONE HOUR RUNS)

OPERATING DATA FOR PCB RUN #1

Elapsed operating time:	70 min. at operating temperature
Feed Rate Total Feed-	3.09 l/min 2.83 kg/min
PCB Feed	0.40 kg/min
Feed Composition (mass)	14.1% PCB 11.0% TCB 74.9% MEK/MeOH
Reactor Operating Temperature	1136°C
Plasma Torch Power	327 kW

Reference: Kolak, et al, 1986.

TABLE 4.7.3.   PCB TEST RESULTS

	Run 1	Run 2	Run 3
Stack Gas Parameters			
Total PCB, (1)	0.013	0.46	3.0
g/dscm* (2)	0.013	0.32	0.011
Total Dioxins, g/dscm	0.076 (3)	0.43	0.13
Total Furans, g/dscm	0.26	1.66	0.30
Total BaP, g/dscm	0.18	0.45	2.8
Scrubber Effluent Parameters			
Total PCB, ppb(1)	1.56	2.15	9.4
(2)	0.06	4.7	0.01
Total Dioxins, ppt	5.8	259	1.35
Total Furans, ppt	1.5	399	1.35
Total BaP, mg/L	0.04	0.92	2.0
Destruction Removal Efficiency			
PCB, Percent DRE			
(1)	>99.99999	99.99994	>99.9999
(2)	99.999999	99.99997	99.999999

(1)   These values are based upon mono-decachlorobiphenyl.
(2)   These values are based upon tri-decachlorobiphenol.
(3)   No tetra or penta dioxins were detected at 0.05 ng on a GL column, except for run #1 where 0.06 ng tetra dioxin was reported.
*g/dscm = grams per dry standard cubic meter

Reference: Kolak, et al., 1986.

Landfills, the electricity cost would amount to $200,000. This estimate is based on the treatment of 330,000 gallons of waste at a rate of 55 gal/hour, a power requirement of 327 kw, and an electricity cost of $0.10 per kilowatt/hr.

Other operating costs include manpower (estimated to be three people), sodium hydroxide for the scrubber, cooling water, and compressed air. One possible cost credit associated with this process is related to the fact that the products of combustion of this process are hydrogen and carbon monoxide. These materials are themselves combustible and could be used as a fuel to run a generator.

4.7.4   Process Status

The construction and testing of the plasma arc system is jointly sponsored by the New York State Department of Environmental Conservation (NYDEC) and the U.S. EPA Hazardous Waste Engineering Research Laboratory (HWERL). The project is comprised of four phases, which are:

Phase 1:   Design and construction of the mobile plasma arc system by the contractor, Pyrolysis Systems, Inc. (PSI).

Phase 2:   Performance testing of the plasma arc system at the Kingston, Ontario, Canada test site.

Phase 3:   Installation of the plasma arc system and additional performance testing at Love Canal, Niagara Falls, N.Y.

Phase 4:   Demonstration testing as designated by NYDEC.

Phase 1 took place in 1982 and Phase 2, the results of which have been presented above, was completed in early 1986. Phase 3 will be initiated later in 1986.

The plasma technology is being jointly marketed by Westinghouse Electric Corporation Waste Technology Services Division and PSI. Once the system has been properly tested, they plan to lease these units to companies or organizations that require the system for waste clean up.

The current system is only designed to handle liquid wastes. Future plans by PSI and Westinghouse include the design of units which could handle contaminated soil and other solid wastes (Haztech News, 1986).

4.8   MOLTEN SALT DESTRUCTION

4.8.1   Process Description

The molten salt destruction process has been under development by
Rockwell International since 1969 (Edwards, 1983). The original intent was to
use the process to gasify coal. A variety of salts can be used, but the most
recent studies have used sodium carbonate (Na_2CO_3) and potassium carbonate
(K_2CO_3) in the 1,450°F to 2,200°F temperature range.

In addition to the Rockwell process, another molten salt process is under
development. The State of New Jersey in late 1982 issued a contract to the
Questex Corporation of New York to evaluate a mobile offsite earth
decontaminator (MOSED), a waste treatment unit based on the molten salt
destruction principle. A status report on the development of this device was
presented at the 1985 Hazpro Conference (Leslie, 1985).

As shown in a schematic of the Rockwell process (Figure 4.8.1), the waste
is fed to the bottom of a vessel containing the liquid salt along with air or
oxygen-enriched air. The molten salt is maintained at an average temperature
of 900°C (1,650°F). The high rate of heat transfer to the waste causes rapid
destruction. Hydrocarbons are oxidized to carbon dioxide and water.
Constituents of the feed such as phosphorous, sulfur, arsenic, and the
halogens react with the salt (i.e., sodium carbonate) to form inorganic salts,
which are retained in the melt. The operating temperatures are low enough to
prevent NO_x emissions (Freeman, 1985; GCA, 1985; Edwards, 1983). Any gases
that are formed are forced to pass through the salt melt before being emitted
from the combustor. If particulates are present in the exhaust gases, a
venturi scrubber or baghouse may be used (GCA, 1985; Edwards, 1983).

Eventually, the build-up of inorganic salts must be removed from the
molten bed to maintain its ability to absorb acidic gases. Additionally, ash
introduced by the waste must be removed to maintain the fluidity of the bed.
Ash concentrations in the melt must be below 20 percent to preserve fluidity
(Edwards, 1983).

Melt removal can be performed continuously or in a batch mode.
Continuous removal is generally used if the waste feed rates are high. The
melt can be quenched in water and the ash can be separated by filtration while

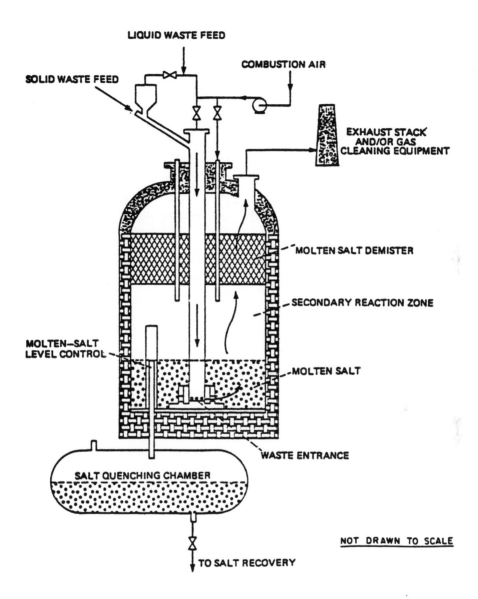

Figure 4.8.1.   Schematic of generalized molten salt incinerator design
[Hitchcock, 1979].

the salt remains in solution. The salt can then be recovered and recycled.
Salt losses, necessary recycle rates, and recycling process design are
strongly dependent on the waste feed characteristics (GCA, 1985; Freeman,
1985; Edwards, 1983).

Restrictive Waste Characteristics--
     The ash content of the melt should be limited to 20 percent in order to
maintain fluidity for a reasonable period of time. The process becomes
inefficient and/or impractical for wastes of high ash content. Also, wastes
with a low water content are destroyed more effectively.

Operating Parameters (Freeman, 1985; GCA, 1985)--
     The following are typical parameters for the molten salt incinerator:

Waste Form	Solid or Liquid Wastes of Low ash and water contents
Operating Temperature	800 to 1000°C (1500° to 1850°F)

Average Residence Time
    Gas Phase
    Solid (or Liquid Phase)
	
	5 seconds
	Hours

Energy Requirements	Fuel to burn waste (if not combustible) Electric power for blowers

4.8.2 Technology Performance Evaluation

     Rockwell International has built two bench scale combustors (0.5 to 2
lb/hr), a pilot plant (55 to 220 lb/hr), and a portable unit (500 lb/hr)
(Edwards, 1983). They have also built a 200 lb/hr coal gasifier based on the
molten salt process.
     Many wastes have been tested in the bench scale unit. Chemical warfare
agents GB, Mustard HD, and VX have been destroyed at efficiencies ranging from
99.999988 to 99.9999995 percent. Other chemicals that have been destroyed
using the molten salt combustion process include: chlordane, malathion, Sevin,
DDT, 2,4-D herbicide, tar, chloroform, perchloroethylene distillation bottoms,
trichloroethane, tributyl phosphate, and PCBs (GCA, 1985; Edwards, 1983).

The PCB trial combustion data are presented in Table 4.8.1.   The destruction efficiency at the lowest operating temperature (1,300°F) exceeded 99.99995 percent. The average residence time of the PCB in the melted salt was 0.25 to 0.50 seconds, based on gas velocities of 1 to 2 ft/sec through the 0.5 ft of melt (GCA, 1985; Edwards, 1983).

Hexachlorobenzene (HCB) and chlordane destruction were tested in the pilot plant facility (Johanson, 1983). Feed rates for HCB and chlordane were as high as 269 lb/hr and 72 lb/hr, respectively. Bed temperatures ranged from 1,685° to 1,805°F, and residence times were close to 2 seconds. HCB destruction efficiencies ranged from nine to eleven nines DRE (99.9999999 to 99.999999999%), and chlordane DREs ranged from seven to eight nines (99.99999 to 99.999999%). The results of the pilot-scale tests are summarized in Table 4.8.2

Smaller scale experiments using 2,3,7,8-TCDD were conducted at the University of Milan in Italy (Bellobono, 1982). A 0.8 in. diameter by 24 in. high reactor was used at temperatures ranging from 1,470°F to 2,190°F. Materials simulating herbicide wastes were prepared by blending 50 percent cellulose powder, 30 percent polyethylene, and selected herbicides. The 2,3,7,8-TCDD concentration was 0.1 to 10 percent by weight. The solids were pulverized to less than 50 mm. 2,3,7,8-TCDD destruction efficiencies ranged from 99.96 percent (at 1,470°F) to 99.98 percent at 2,190°F.

4.8.3   Costs of Treatment

The molten salt process has not been developed to the scale that specific cost projections can be developed. However, the equipment cost for the 200 lb/hr capacity pilot unit has been estimated to be $1.4 million (GCA, 1984).

The most significant factor that affects the cost of molten salt incineration is the frequency with which the salt bed needs to be replaced. For a high chlorine content waste, the replacement of the bed can be as high as one pound of bed for one pound of waste destroyed (GCA, 1984).

Costs will be higher for wastes which require auxiliary fuel to sustain combustion in the molten salt bed. Wastes with heating values ≥4000 BTU/lb should not require supplemental fuel (GCA, 1984).

TABLE 4.8.1.   PCB COMBUSTION TESTS IN SODIUM-POTASSIUM-CHLORIDE-CARBONATE
MELTS [Edwards, 1983]

Temp (°F)	Stochiometric air (%)	Concentration of KCl, NaCl in melt (wt %)	Extent of PCB destruction[a] (%)	Concentration of PCB in off-gas[a] (g/m³)
1598	145	60	>99.99995	52
1526	115	74	>99.99995	65
1292	160	97	>99.99995	51
1643	180	100	>99.99993	59
1427	125	100	>99.99996	44
1427	90	100	>99.99996	66

[a]PCBs were not detected in the off-gas, i.e., values shown are detection
limits.

Reference: GCA, 1985; Edwards, 1983.

TABLE 4.8.2.   SUMMARY OF PILOT-SCALE TEST RESULTS

	PCB	Chlordane
Combustor Feed Rate (lb/hr)	20.9 - 122.0	12.1 - 32.7
Combustor Off-gas		
- mg/m^3	2.7 x 10^{-4} - 7.1 x 10^{-2}	5.3 x 10^{-3} - 6.8 x 10^{-2}
- ppmv	2.3 x 10^{-5} - 6.1 x 10^{-3}	3.2 x 10^{-4} - 4.1 x 10^{-3}
Baghouse		
- mg/m^3	<6 x 10^{-6} - 1.6 x 10^{-4}	<3.6 x 10^{-4} - 4.4 x 10^{-3}
- ppmv	<5.2 x 10^{-7} - 1.4 x 10^{-5}	<2.1 x 10^{-5} - 2.6 x 10^{-4}
Spent Melt (ppmv)	0.001 - 0.104	0.0044 - 1.2
NO$_x$ (ppmv)	70 - 125	0.5 - 630
HC (ppmv)	35 - 110	0.4 - 60
Particulate (mg/m^3)	<6.2 x 10^{-3} - 0.107	4.1 x 10^{-3} - 1.75 x 10^{-2}
DRE (%)	11-9's - 9-9's	8-9's - 7-9's

Note: The pH of the liquid in a small sampling scrubber in the off-gas line remained basic throughout the test indicating essentially no HCl emission.

Reference: Freeman, 1985.

4.8.4   Process Status

Rockwell has constructed several molten salt units of varying sizes. The company has conducted extensive tests in two sizes of units; bench-scale combustors for feeds of up to 10 lb/hr wastes, and a pilot-scale unit for feeds of up to 250 lb/hr of wastes (Freeman, 1985).

Several developments will be needed if molten salt combustion is to be applied to dioxin-contaminated wastes. The 99.9999 percent DRE required by the current RCRA regulations must be demonstrated for dioxins, and additional testing with dioxin-contaminated wastes (both liquids and solids) needs to be performed on a larger scale. Research to develop more economical construction materials may also be required.

As indicated above, molten salt combustion is not currently practical for the treatment of dioxin-contaminated wastes. Additional research and development is required, but Rockwell has no plans for such further activity. The status of the New Jersey MOSED unit is not known.

4.9 SUPERCRITICAL WATER OXIDATION

4.9.1 Process Description

The supercritical water (SCW) oxidation process utilizes the properties of water at pressures greater than 218 atmospheres combined with temperatures above 374°C to effect oxidation of organics such as TCDD (Thomason and Modell, 1984; Josephsory, 1982). Above these temperatures and pressures, water is in its supercritical state and exhibits solubility characteristics which are the inverse of normal liquid water properties (Thomason and Modell, 1984; Josephsory, 1982). Thus, organics become almost completely soluble and inorganic salts become only sparingly soluble and tend to precipitate.

The steps involved in the SCW oxidation process (as developed by Modar, Inc.) are diagrammed in Figure 4.9.1. Initially, the waste (in the form of an aqueous solution or slurry) is pressurized and heated to supercritical conditions by mixing it with recycled reactor effluent (Modar, 1982; Modell, 1984). Compressed air is also mixed with the feed to serve as source of oxygen for the reactions. Oxygen and air are miscible with water under supercritical conditions, thereby enabling the homogeneous operation of the process. The homogenized mixture is then pumped to the oxidizer where organics are rapidly (residence times average 1 minute) oxidized. Oxidation is achieved under homogeneous conditions (single-phase supercritical fluid) and therefore higher effective oxygen concentrations and destruction efficiencies can be achieved with shorter residence times than with other similar processes (i.e., the wet oxidation process).

The release of combustion heat from the oxidation reactions causes temperatures in the oxidizer reactor to rise to 1112 to 1202°F (Modell, 1984; Freeman, 1985; GCA, 1985; GCA, 1984). The reactor effluent then enters a cyclone (solids separator) where inorganic salts are precipitated out (at temperatures above 450°C) (Modell, 1984; GCA, 1985; GCA, 1984; Freeman, 1985). The fluid effluent of the solid separator consists of superheated, supercritical water, nitrogen, and carbon dioxide. A portion of the superheated, supercritical water is directed to an eductor so that it can be recycled to heat the incoming waste feed (initial step in the process). Modar, Inc. suggests that the remaining effluent, which consists of a high temperature, high pressure fluid, can be cooled to subcritical temperatures in

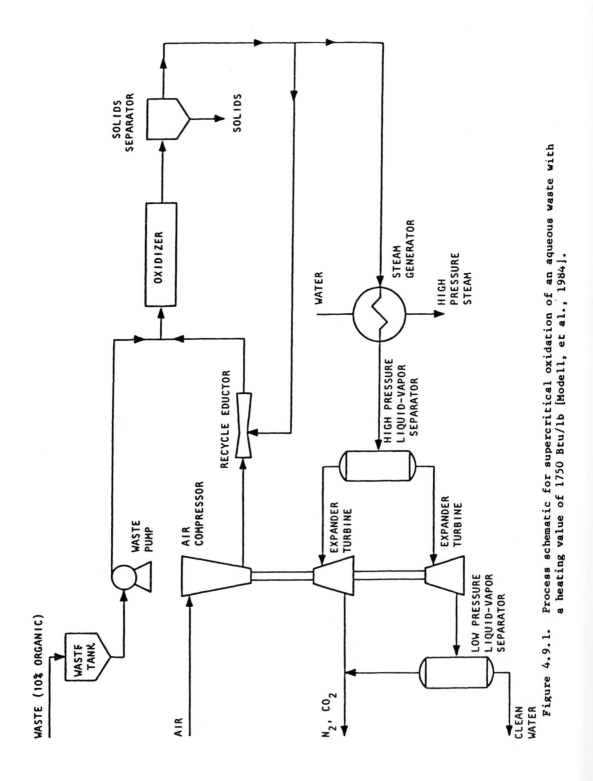

Figure 4.9.1.   Process schematic for supercritical oxidation of an aqueous waste with a heating value of 1750 Btu/lb [Modell, et al., 1984].

a heat exchanger and the resulting steam can be used with turbines to generate energy (Modar, 1982; Modell, 1984). However, the cost-effectiveness of the turbine power generation system is limited to certain cases.

The supercritical oxidation process results in conversion of carbon and hydrogen compounds from the organic compound to CO_2 and H_2O (Swanson, et al., 1984; Josephson, 1982). Chlorine atoms are converted to chloride ions and can be precipitated as sodium chloride with the addition of basic materials to the feed. Gaseous emissions consist primarily of carbon dioxide with smaller amounts of oxygen and nitrogen gas, which do not require auxilliary treatment for offgases. Solid emissions consist of precipitated inorganic salts (chlorine produces chloride salts, nitro compounds precipitate as nitrates, sulfur compounds as sulfates, and phosphorous compounds as phosphates). The liquid effluent consists of a purified water stream, which can be used for process water.

Restrictive Waste Characteristics (Thomason and Modell, 1984; GCA, 1985; GCA, 1984; Freeman, 1985)--
    Certain restrictions exist concerning the types of waste that can be treated using the supercritical water oxidation system. These restrictions are:

1. Organic concentrations need to be less than 20 percent by weight in order for the process to be cost-effective; higher concentrations can be diluted by mixing with dilute wastewater or with pure water.

2. The waste needs to be in the form of an aqueous solution or slurry. Solids can be mixed with water to form a slurry.

3. Costs are higher if the waste has a fuel value of greater than 1750 Btu/lb, a value equivalent to that exhibited by a waste consisting of 10 percent by weight of benzene or its equivalent. This is the optimal heat for achieving a reactor exit temperature of 600 to 650°C. Wastes with greater than a 10 percent benzene equivalent should be diluted, and fuel should be added to wastes with less than a 10 percent benzene-equivalent.

Operating Parameters (Sieber, 1986)--
    The following are typical operating parameters for the SCW system:

Waste Form	Aqueous solution or slurry of organics
Temperature	450 to 650°C
Pressure	220 to 250 atmospheres
Average Residence Time	less than 1 minute
Feed rate	-

4.9.2 Technology Performance Evaluation

Modar has built and tested bench scale supercritical water reactors for destruction of urea, chlorinated organics, and dioxin-containing wastes. Skid-mounted, transportable systems with a capacity of 50 gal/day have been designed as well as larger-scale stationary units.

A reactor, constructed of Iconel 628 and measuring 19.6 inches long with an inside diameter of 0.88 inches, was used to investigate urea destruction (GCA, 1985; GCA, 1984). Additional tests of the supercritical water oxidation process were conducted using a similarly constructed Hastelloy C-276 reactor with dimensions of 24 inches in length and 0.88 inch inside diameter (GCA, 1985; GCA, 1984). Tests on chlorinated organics were performed with this reactor. Table 4.9.1 summarizes the compositions of the various waste feeds used in the test runs. Liquid influents and effluents were analyzed for total organic carbon TOC by GC/MS (Modell, 1982). Gaseous effluents were analyzed by GC for low molecular weight hydrocarbons (Modell, 1982). The results of these analyses and the calculated destruction and removal efficiencies (DREs) are shown in Table 4.9.2. Chlorinated dibenzo-p-dioxins were searched for specifically, but none were found in the effluents.

A laboratory-scale trial burn was conducted using a feed consisting of synthesized dioxin added to trichlorobenzene (at 100 ppm concentrations) (Killiley, 1986). According to Modar, the process achieved greater than six nine's DRE based on the analytical detection limits for gas and liquid effluents (Killiley, 1986). Modar has performed studies on dioxin-contaminated soils for private clients, including a field demonstration using their pilot scale unit for the New York Environmental Conservation Department (Killiley, 1986). Although the results of these tests are not available for release at this time, the DREs for TCDD were reportedly greater than 6 nines (Killiley, 1986).

4.9.3 Costs of Treatment

The most significant operating cost factor is the cost of oxygen consumed (GCA, 1985). Although compressed air can be used as the source of oxygen, the cost of power as well as the high capital cost of appropriate compressors has

TABLE 4.9.1.   COMPOSITION OF FEED MIXTURES FOR TEST RUNS [Modell, 1982]

		Wt %	Wt % Cl
Run 11			
DDT	$C_{14}H_9Cl_5$	4.32	2.133
MEK	C_4H_8O	95.68	--
		100.0	2.133
Run 12			
1,1,1-trichloroethane	$C_2H_3Cl_3$	1.01	0.806
1,2-ethylene dichloride	$C_2H_2Cl_2$	1.01	0.739
1,1,2,2-tetrachloroethylene	C_2Cl_4	1.01	0.866
o-chlorotoluene	C_7H_7Cl	1.01	0.282
1,2,4-trichlorobenzene	$C_6H_3Cl_3$	1.01	0.591
biphenyl	$C_{12}H_{10}$	1.01	--
o-xylene	C_8H_{10}	5.44	--
MEK	C_4H_8O	88.48	--
		100.0	3.284
Run 13			
hexachlorocyclohexane	$C_6H_6C_6$	0.69	0.497
DDT	$C_{14}H_9Cl_6$	1.00	0.493
4,4'-dichlorobiphenyl	$C_{12}H_8Cl$	1.57	0.495
hexachlorocyclopentadiene	C_6Cl_6	0.65	0.505
MEK	C_4H_8O	96.09	--
		100.0	1.99
Run 14			
PCB 1242	$C_{12}H_xCl_{4-6}$	0.34	0.14
PCB 1254	$C_{12}H_xCl_{5-8}$	2.41	1.30
Transformer oil	$C_{10}-C_{14}$	29.26	--
MEK	C_4H_8O	67.99	--
		100.0	1.44
Run 15			
4,4-'dichlorobiphenyl	$C_{12}H_8Cl$	3.02	.96
MEK	C_4H_8O	96.98	--
		100.0	0.96

TABLE 4.9.2.   SUMMARY OF RESULTS: OXIDATION OF ORGANIC CHLORIDES
[Modell, 1982]

Run No.	11	12	13	14	15
Residence time (min)	1.1	1.1	1.1	1.1	1.3
Carbon analysis					
Organic carbon in (ppm)	26,700.	25,700.	24,500.	38,500.	33,400.
Organic carbon out (ppm)	2.0	1.0	6.4	3.5	9.4
Destruction efficiency (%)	99.993	99.996	99.975	99 .991	99.97
Combustion efficiency (%)	100.	100.	100.	100.	100.
Gas composition					
O_2	25.58	32.84	37.10	10.55	19.00
CO_2	59.02	51.03	46.86	70.89	70.20
CH_4	—	--	--	--	--
H_2	—	--	--	--	--
CO	--	--	--	--	--
Chloride analysis					
Organic chloride in (ppm)	876.	1266.	748.	775.	481.
Organic chloride out (ppm)	.023	.037	<.028	.032	.036
Organic chloride conversion (%)	99.997	99.997	99.996	99.996	99.993
GC/MS effluent analysis					
Compound B[a] (ppb Cl)	—	--	--	—	--
C	--	--	--	--	--
E	—	9	--	14	--
F	18	12	18	--	--
H	—	--	<4.	--	--
K	5	16	<5.	6	--
M	—	--	0.2	--	--
N	--	--	0.3	--	36
O	—	--	--	12	--

[a]Compounds searched by GC/MS Analysis

Notes: p-isomers are assumed, based on
the position of the chlorine atoms
in the starting materials.

Compound F — No authentic MS
Compound G — No MS available in literature

led Modar to use liquefied oxygen as the primary oxygen source. Oxygen demand and heat content of an organic waste are usually directly related. Therefore, the heating value of the waste and waste throughput can be used to make a preliminary estimate of treatment costs.

Table 4.9.3 presents waste treatment costs based on an aqueous waste with a 10 percent by weight benzene-equivalent and a heat content of 1,800 Btu/lb. This is the optimal heat content of a cold feed for this process to attain a reactor exit temperature of 600 to 650°C (GCA, 1984). Other factors on which the costs in Table 4.9.3 are based are:

(1)   the system is installed at the site of the waste generator;

(2)   the units are owned and operated by the waste disposer; and

(3)   the units are not equipped with power recovery turbines.

If the waste has a fuel value of greater than 1,800 Btu/lb, the cost will be higher per unit of waste processed (GCA, 1984). In treating a waste with a higher organic content, it is recommended that the waste is diluted to a 10 percent benzene-equivalent (Modell, 1984; GCA, 1984). Therefore, the increase in cost will be in proportion to the increase in organic content.

If the waste has a heat content of between 5 and 10 percent benzene-equivalent, fuel can be added to the waste to bring the heat content up to 10 percent benzene-equivalent without appreciable cost increases (GCA, 1984: Modell, 1984). If, however, the waste is dilute (2 to 3 percent benzene equivalent) it is more economical to use a combination of fuel with regenerative heat exchange (GCA, 1984: Modell, 1984).

4.9.4   Process Status

Design of full scale units for supercritical fluid oxidations is underway at Modar, Inc (Killiley, 1986). Commercial units should be available in 1987 although their cost effectiveness for dioxin wastes has not been established. Treatment of contaminated soils, without pretreatment to extract adsorbed dioxin, would appear to be impractical.

TABLE 4.9.3.   MODAR TREATMENT COSTS FOR ORGANIC
CONTAMINATED AQUEOUS WASTES[a]

Waste capacity		Processing cost	
gal/day	ton/day	$/gal[b]	$/ton[b]
5,000	20	$0.75 - $2.00	$180 - $480
10,000	40	$0.50 - $0.90	$120 - $216
20,000	80	$0.36 - $0.62	$ 86 - $149
30,000	120	$0.32 - $0.58	$ 77 - $139

[a]Based upon an aqueous waste with 1800 Btu/lb heating
value and inorganic solids of between 1% and 10%.

[b]Does not include energy recovery value of approximately
$0.05 per gallon.

Source: Sieber, 1986.

4.10   IN SITU VITRIFICATION

4.10.1   Process Description/Flow Diagram

The basic principle of the In Situ Vitrification (ISV) process involves placement of electrodes into a contaminated soil zone and then passing electrical current between the electrodes. This "joule heating" principle, which utilizes the soil as the resistance element in an electric circuit, creates temperatures in excess of 1,350°C and leads to the melting of the soil and subsequent formation of a stable/immobile molten glass or crystalline substance.

The basic ISV process was developed by Battelle Memorial Institute's, Pacific Northwest Laboratories (PNL) under a funding program with the U.S. Department of Energy's (DOE) Richland Operations Office. This emerging technology was developed as a potential method for the in place stabilization of transuranic (TRU) contaminated materials (Fitzpatrick, et.al., 1984). However, recently this concept has been envisioned as an applicable, in situ treatment technology for contaminated soils at hazardous waste sites. The overall process development and application to dioxin-contaminated soils is described in the following narrative (Brouns, et.al., 1982; Fitzpatrick, et.al., 1984). Figure 4.10.1 provides a conceptual schematic diagram of a proposed ISV system.

The first step in the use of the ISV process requires that the boundaries encompassing the area to be treated be clearly identified. Once this condition is satisfied, molybdenum or graphite electrodes are then inserted into the contaminated soil area at the four corners of the boundary to form a square. A high voltage, over 4,000 volts for a large scale vitrification process, is then applied until a vitreous soil mass is produced.

During the melting process, organic materials tend to pyrolyze, rise to the surface of the molten glass, and combust when brought into contact with air. Other components such as fission products, transuranics, heavy metals, and non-volatile organics become trapped in the molten soil product. The volatile organic combustion products are collected and treated to prevent the transfer of pollutants from soil to air. A hood is placed over the area being vitrified to perform three functions:

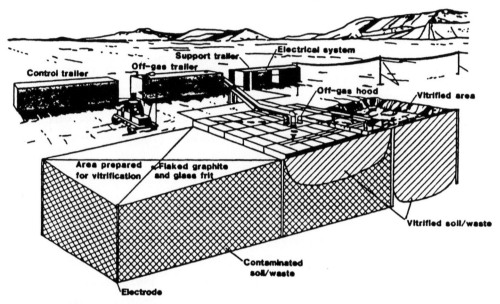

Source: Adapted from Pacific Northwest Laboratory

Figure 4.10.1.   Schematic diagram of an in situ vitrification operation
[Hazardous Waste Consultant, 1985].

(1) to collect the gas products of the combustion reaction;

(2) to act as a chamber for combustion of the pyrolyzed volatile
     organics; and

(3) to support the electrodes placed in the soil.

The off-gas treatment system consists of three stages. First, the
off-gas is cooled and scrubbed in a quencher and tandem nozzle scrubber to
remove larger particles. The water in the gas is then removed by a vane
separator followed by a condenser and a second vane separator. Finally, the
gas is heated above its dewpoint to maintain an unsaturated gas stream which
is then filtered by high efficiency particulate air (HEPA) filters. Following
this stage the stream is discharged through a stack. A schematic of this
off-gas treatment system is illustrated in Figure 4.10.2.

4.10.2 Technology Performance Evaluation

Since the initiation of this program with the DOE, the PNL has conducted
several engineering-scale and pilot-scale tests with the ISV process on
radionuclide wastes. Specifically, 21 engineering-scale (laboratory) tests
have been conducted which produced a vitrified mass of between 0.05 and 1.0
tons per test at a power level of 30 kW. Pilot-scale (field) tests have also
resulted in up to 10 tons of vitrified mass in each of the seven tests.

Additionally, performance studies have been conducted by PNL on four
different aspects of ISV (Buelt, et al., 1984). The effect of variations in
soil types was studied to determine the scope of the potential market.
Examining the quality of the vitrified waste helped to qualify this same
market, since determining the behavior of organic hazardous wastes during ISV
processing could extend this market into areas such as the treatment of dioxin
contaminated soil.

PNL have conducted experiments on nine different kinds of radioactive
soils from all over the United States. These tests proved there is no
expected degradation (as measured by variations in properties and as
electrical and thermal conductivities, fusion, temperature, viscosity, and
chemical composition) of the ISV systems capabilities due to varying soil
types. However, ISV has only been conducted on Hanford nuclear waste soils so
conclusive data has not yet been generated.

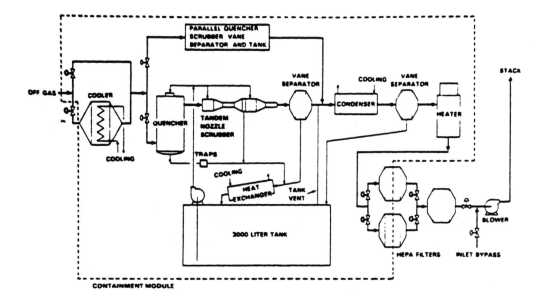

Figure 4.10.2.   Schematic of large-scale off-gas
                  treatment system (Fitzpatrick, 1984).

Further analysis of the vitrified wastes indicates that the waste blocks are expected to maintain their integrity for more than 10,000 years (Oma, et al., 1983). This vitrified waste has also been shown to exhibit leach resistance superior to marble and possess durability similar to granite (Strachan, et al., 1980 and MCC, 1981). It has been suggested, that for a vitrified site with engineered barriers exposure could be reduced by a factor of 10^5 for TRU wastes (Buelt, et al., 1984). The reduction in exposure obtained for toxic and heavy metal wastes is currently unknown.

Experimental work on the treatment of organic and hazardous wastes (e.g., Co, Cd, carbon tetrachloride, and dichlorobenzene) has resulted in the following three conclusions (Oma, et al., 1984). First, it is apparent that gaseous releases of combustible organics result in a higher release fraction when implementing ISV. Secondly, certain organics are pyrolyzed in the soil and consequently achieve complete combustion in the off-gas hood maintained over the site being vitrified. Finally, burying the wastes to a greater depth reduces the potential for release of hazardous elements.

While these conclusions represent distinct advantages of the system, the process also has several limitations. The most severe limitation restricts the process to vitrification depths of approximately 13 meters (40 feet)(Buelt, J., 1986). The electrical nature of the system poses another problem, especially as it relates to soil moisture. Additionally, since the process depends on the conductivity of the medium being vitrified, metallic conductors in the soil can reduce efficiency. The potential even exists for metal pipes or bars to short out the electrodes or for sealed containers housing highly combustible organics to explode. Insulators and void volumes (such as empty plastic containers) can also seriously affect performance. It is possible to get collapse in the molten block being formed if the void volume is greater than the volume of molten soil already formed. Some pretreatment of the soil may be necessary to minimize this problem.

4.10.3 Costs of Treatment

The cost considerations reported by PNL, and discussed below for TRU wastes treated by the ISV process, account for charges associated with site preparation, consumable supplies such as electrical power, and operational

costs such as labor and annual equipment charges (Oma, et al., 1983).
Specifically, for variations in manpower levels, power source costs, and
degree of heat loss it was determined that the costs for TRU waste
vitrification ranges from 160 to 360 $/m^3 to vitrify to a depth of
5 meters. These costs are a function of many variables but are most sensitive
to variations in the amount of moisture in the soil and the cost of electrical
power in the vicinity of the process. Figure 4.10.3, developed by PNL,
illustrates the variation in total costs as a function of both electrical
power costs and moisture content of TRU soil experimentally treated. The
vertical line represents the value beyond which it is more cost effective to
lease a portable generator.

Recently, PNL has assessed the cost implications for ISV treatment of
three additional waste categories; i.e., industrial sludges and hazardous
waste (PCB) contaminated soils at both high and low moisture contents (Buelt,
J., 1986). Representatives at PNL indicated that for industrial sludges with
moisture contents of 55 to 75 percent (classified as a slurry), the total
costs would range from 70 to 130 $/m^3. Treatment of high (greater than 25
percent) moisture content hazardous waste-PCB contaminated soil would cost
approximately 150 to 250 $/m^3 versus costs of 128 to 230 $/m^3 for low
(approximately 5 percent) moisture content PCB contaminated soil.

As these recent data and past TRU waste cost data suggest, the moisture
content of the contaminated material is particularly important in influencing
treatment costs. High moisture content increases both the energy and length
of time required to treat the contaminated material. Furthermore, PNL
representatives suggest that treatment costs are also influenced by the degree
of off-gas treatment required for a given contaminated material (i.e., ISV
application to hazardous chemical wastes will likely not require as
sophisticated an off-gas treatment system as would TRU waste treatment).

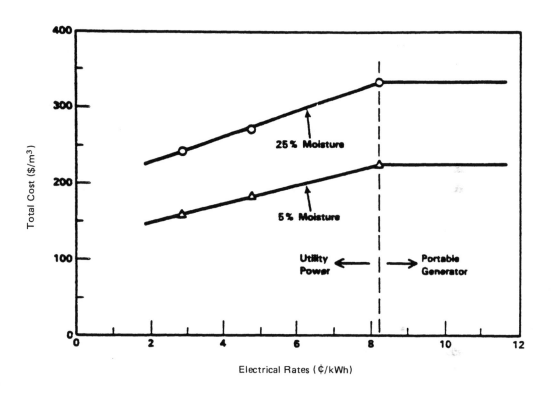

Figure 4.10.3. Cost of insitu vitrification for Transuranic wastes as a function of electrical rates and soil moisture (Fitzpatrick, 1984).

4.10.4   Process Status

   As briefly indicated above in the "Cost" discussions, PNL has recently
assessed the treatment and costs associated with hazardous waste contaminated
soils (Buelt, J., 1986). During the summer of 1985, tests were conducted for
the Electric Power Research Institute (EPRI) on PCB contaminated soil. Note
that while the draft report on these tests has been completed, it has not been
published and/or made available to date. However, an EPRI project summary
publication, dated March 1986, entitled "Proceedings: 1985 EPRI PCB Seminar"
(EPRI CS/EA/EL 4480) has recently been made available to EPRI members.
Preliminary results suggest that a destruction/removal efficiency (DRE) of six
to nine nines was achieved from the off-gas treatment system overall, and that
a vitrification depth of 2 feet was achieved. Additional information will
soon be available to the public. PNL expects to continue with research in the
area of hazardous waste soils.

REFERENCES

Ackerman, D.G., L.L. Scinto, C.C. Shih, and B.J. Matthews. TRW, Inc. The Capability of Oceanic Incineration - A Critical Review and Rebuttal of the Kleppinger Report. May 1983.

Ackerman, D.G. Sitex Consultants East, Inc. Draft Final Report: The Capability of Oceanic Incineration - A Critical Review of the Kleppinger-Bond Report. March 1986.

Bailey, William. Rollins Environmental Services, Inc. Telephone Conversation with Lisa Farrell, GCA Technology Division, Inc. Re: PCBs Incineration Costs. May 13, 1986.

Barner, H.E., J.S. Chartier, H. Beisswenger, and H.W. Schmidt. Lurgi Corporation. Application of Circulating Fluid Bed Technology to the Combustion of Waste Materials. Environmental Progress, 4(2): 125-130. May 1985.

Barton, Thomas G. Mobile Plasma Pyrolysis. Hazardous Waste, 1(2): 237-247. 1984.

Bellobono, I.R., E. Selli, and L. Veronese. Destruction of Dichloro- and Trichloro-phenoxyacetic acid esters containing 2,3,7,8-Tetrachlorodibenzo-p-dioxin by Molten Salt Combustion Technique. Acqua-Aria. January 1982.

Bond, Desmond H. At-Sea Incineration of Hazardous Wastes. Environmental Science & Technology, 18(5): 148A-152A. May 1984.

Bonner, T.A., et al. Engineering Handbook for Hazardous Waste Incineration. Report prepared for U.S. EPA, Cincinnati, Ohio. SW-899. June 1981.

Boyd, J., H.D. Williams, and T.L. Stoddard. Destruction of Dioxin Contamination by Advanced Electric Reactor. Preprinted Extended Abstract of Paper Presented Before the Division of Environmental Chemistry, American Chemical Society, 191st National Meeting, New York, New York: Vol 26, No. 1. April 13-18, 1986.

Boyd, James. J.M. Huber Corporation. Telephone Conversations with Lisa Farrell, GCA Technology Division, Inc. January 28, 1986; April 3, 1986; May 1, 1986.

Brouns, R.A., and C.L. Timmerman. Pacific Northwest Laboratories, Richland, Washington. In Situ Thermoelectric Stabilization of Radioactive Wastes. In: Proceedings of the Waste Management 1982 Meeting in Tucson, Arizona. PNL-SA-9924. 1982.

Brouns, R.A., J.L. Buelt, and W.F. Bonner. In Situ Vitrification of Soil. U.S. Patent 4, 376, 598. 1983.

Brown, William. Chemical Waste Management, Inc. Telephone Conversation with Lisa Farrell, GCA Technology Division, Inc. March 28, 1986.

Buelt, J.L., et al. Battelle Memorial Institute, Pacific Northwest Laboratories, Richland, Washington. An Innovative Electrical Technique for In-Place Stabilization of Contaminated Soils. In: Proceedings of the American Institute of Chemical Engineers 1984 Summer Meeting in Philadelphia, Pennsylvania. 1984.

Buelt, J.L. Battelle Memorial Institute, Pacific Northwest Laboratories, Richland, Washington. Telephone Conversation with Michael Jasinski, GCA Technology Division, Inc. 1986.

California State Air Resources Board. Technologies for the Treatment and Destruction of Organic Wastes as Alternatives to Land Disposal. 1982.

Carnes, Richard A., and Frank C. Whitmore. Characterization of the Rotary Kiln Incinerator System at the U.S. EPA Combustion Research Facility (CRF). Hazardous Waste, 1(2): 225-236. 1984.

Carnes, Richard A. U.S. EPA, Combustion Research Facility. U.S. EPA Combustion Research Facility Permit Compliance Test Burn. In: Proceedings of the Eleventh Annual Research Symposium on Incineration and Treatment of Hazardous Waste, sponsored by U.S. EPA-HWERL. Cincinnati, Ohio, April 29-May 1, 1985. EPA/600/9-85/028. September 1985.

Carnes, Richard A. U.S. EPA, Combustion Research Facility. Telephone Conversation with Lisa Farrell, GCA Technology Division, Inc. January 29, 1986.

Chang, Daniel P.Y., and Nelson W. Sorbo. University of California, Department of Civil Engineering. Evaluation of a Pilot Scale Circulating Bed Combustor with a Surrogate Hazardous Waste Mixture. In: Proceedings of the Eleventh Annual Research Symposium on Incineration and Treatment of Hazardous Waste, sponsored by U.S. EPA-HWERL. Cincinnati, Ohio, April 29-May 1, 1985. EPA/600/9-85/028. September 1985.

Chemical Engineering. New Units Give Boost to Sludge Incineration. July 9, 1984.

Clark, W.D., J.F. La Fond, D.K. Moyeda, W.F. Richter, W.R. Seeker, and C.C. Lee. Engineering Analysis of Hazardous Waste Incineration; Failure Mode Analysis for Two Pilot Scale Incinerators. In: Proceedings of the Eleventh Annual Research Symposium on Incineration and Treatment of Hazardous Waste, sponsored by U.S. EPA-HWERL. Cincinnati, Ohio, April 29-May 1, 1985. EPA/600/9-85/028. September 1985.

Daily, Philip L. Shirco Infrared Systems, Inc. Performance Assessment of Portable Infrared Incinerator. Storage & Disposal. 1985.

Daily, Philip L. Shirco Infrared Systems, Inc. Performance Assessment of
     Portable Infrared Incinerator. Preprinted Extended Abstract of Paper
     Presented Before the Division of Environmental Chemistry, American
     Chemical Society, 191st National Meeting. New York, New York: Vol. 26,
     No. 1. April 13-18, 1986.

Eaton, H.C., M.E. Tittlebaum, and F.K. Cartledge. Louisiana State
     University. Techniques for Microscopic Studies of Solidification
     Technologies. In: Proceedings of the 11th Annual Research Symposium on
     Incineration and Treatment of Hazardous Waste, sponsored by U.S.
     EPA-HWERL. Cincinnati, Ohio, April 29-May 1, 1985. EPA/600/9-85/028.
     September 1985.

Edwards, B.H., J.N. Paullin, K.C. Jordan. Noyes Data Corporation, Park Ridge,
     New Jersey. Emerging Technologies for the Control of Hazardous Wastes.
     1983.

Ellis, William D., William H. Vick, Donald E. Sanning, and Edward J. Opatkin.
     Evaluation of Stabilized Dioxin Contaminated Soils. In: Proceedings of
     the EPA-HWERL 11th Annual Research Symposium, Cincinnati, Ohio.
     April 29-May 1, 1985.

Ellis, William. JRB Associates. Telephone Conversation with Lisa Farrell,
     GCA Technology Division, Inc. May 15, 1986.

Environmental Research and Technology, Inc. Final Report: Onsite
     Incineration Testing of Shirco Infrared Systems Portable Pilot Test Unit,
     Times Beach Dioxin Research Facility, Times Beach, Missouri. Prepared
     for Shirco Infrared Systems, Inc. Report No. 815-85-2.
     November 14, 1985.

Fitzpatrick, V.F., et al. In Situ Vitrification - A Potential Remedial Action
     Technique for Hazardous Wastes. In: Proceedings of the 5th National
     Conference on Management of Uncontrolled Hazardous Waste Sites,
     Washington, D.C. 1984.

Flax, Louis. LOPAT Enterprises, Inc. Telephone Conversation with Lisa
     Farrell, GCA Technology Division, Inc. May 19, 1986.

Freeman, Harry M. Hazardous Waste Destruction Processes. Environmental
     Progress. Volume 2, Number 4. November 1983.

Freeman, Harry M. U.S. EPA, Hazardous Waste Engineering Research Laboratory,
     Thermal Destruction Branch. Project Summary: Innovative Thermal
     Hazardous Waste Treatment Processes. 1985.

Freeman, Harry M., and Robert A. Olexsey. A Review of Treatment Alternatives
     for Dioxin Wastes. Journal of the Air Pollution Control
     Association (JAPCA), 36(1): 66-75. January 1986.

Freestone, F., R. Miller, and C. Pfrommer. Evaluation of Onsite Incineration for Cleanup of Dioxin-Contaminated Materials. In: Proceedings of the International Conference on New Frontiers for Hazardous Waste Management, sponsored by: U.S. EPA-HWERL, NUS Corporation, National Science Foundation, and American Academy of Environmental Engineers. Pittsburgh, Pennsylvania, September 15-18, 1985. EPA/600/9-85/025. September 1985.

Freestone, Frank. U.S. EPA-HWERL, Edison, New Jersey. Telephone Conversation with Lisa Wilk, GCA Technology Division, Inc. August 5, 1986.

GA Technologies, Inc. Brochure: Circulating Bed Waste Incineration. 1984.

GCA Technology Division. Screening to Determine the Need for Standards of Performance for Industrial and Commercial Incinerators. Prepared for the U.S. EPA, Office of Air Quality Planning and Standards, under EPA Contract Nos. 68-02-2607 and 68-02-3057. January 1979.

GCA Technology Division. Draft Final Report: Technology Overview - Circulating Fluidized Bed Combustion. Prepared for U.S. EPA, Office of Research and Development, under EPA Contract No. 68-02-2693, GCA-TR-81-91-G. August 1981.

GCA Technology Division. Utilization of Non-Land Disposal Alternatives to Handle Superfund Wastes. Prepared for the U.S. EPA, Office of Solid Waste, Waste Management and Economics Division. July 25, 1984a.

GCA Technology Division. Final Report: Technical Assessment of Treatment Alternatives for Wastes Containing Halogenated Organics. Prepared for U.S. Environmental Protection Agency, Office of Solid Waste, Waste Treatment Branch, under EPA Contract No. 68-01-6871, Work Assignment No. 9. GCA-TR-84-149-G. October 1984b.

GCA Technology Division. Detailed Review Draft Report: Identification of Remedial Technologies. Prepared for U.S. EPA, Office of Waste Programs Enforcement, under EPA Contract No. 68-01-6769, Work Assignment No. 84-120. GCA-TR-84-109-G-(0). March 1985.

Gregory, R.C. Rollins Environmental Services, Inc. Design of Hazardous Waste Incinerators. Chemical Engineering Progress. April 1981.

Hazardous Waste Consultant. Stabilizing Organic Wastes: How Predictable Are The Results? Volume 3, Issue 3, Pages 1-18 to 1-19. May/June 1985.

Hazardous Materials Intelligence Report. EPA Incineration Test, Public Hearing on AL Waste Law Amendments. December 6, 1985a.

Hazardous Materials Intelligence Report. EPA to Issue Permits for Ocean Incineration Tests. December 6, 1985b.

Hazardous Materials Intelligence Report. Texas Company Markets Transportable Infrared Incinerator. January 10, 1986a.

Hazardous Materials Intelligence Report. First Commercial Dioxin Incineration Permit Granted to J.M. Huber. January 24, 1986b.

Hazardous Materials Intelligence Report. Opposition Raised to EPA's Ocean Incineration Proposal. 7(6): 2-3. February 7, 1986c.

Hazardous Materials Intelligence Report. EPA's Ocean Incineration Proposal Delayed by NOAA. February 28, 1986d.

Hazardous Waste Consultant. Volume 3, Issue 1, Pages 4-4 and 4-5. McCoy & Associates Publication. January/February 1985.

Hazel, Ralph. U.S. EPA, Region VII. Telephone Conversation with Lisa Wilk, GCA Technology Division, Inc. August 4, 1986.

Haztech Notes. Plasma Arc Technology Used to Atomize Liquid Organics. 1(5): 33-34. 1986.

Hicks, James. ENSCO, Inc. Telephone Conversation with Lisa Farrell, GCA Technology Division, Inc. February 5, 1986.

Hill, Michael. Shirco Infrared Systems, Inc. Dallas, Texas. Telephone Conversation with Lisa Farrell, GCA Technology Division, Inc. January 14, 1986.

Hitchcock, D.A. Solid Waste Disposal: Incineration. Chemical Engineering, 86(11): 185-194. May 21, 1979.

Horning, A.W., and H. Masters. Rockwell International, Newbury Park, California. Destruction of PCB-Contaminated Soils With a High-Temperature Fluid-Wall (HTFW) Reactor. Prepared for U.S. EPA, Office of Research and Development, Municipal Environmental Research Laboratory, Cincinnati, Ohio. EPA-600/D-84-072. 1984.

IT Corporation. Interim Summary Report on Evaluation of Soils Washing and Incineration As Onsite Treatment Systems for Dioxin-Contaminated Materials. Prepared for U.S. EPA, Hazardous Waste Engineering Research Laboratory, under EPA Contract No. 68-03-3069. June 7, 1985a.

IT Corporation. Dioxin Trial Burn Data Package, EPA Mobile Incineration System at the James Denney Farm Site, McDowell, Missouri. Prepared for U.S. EPA, Hazardous Waste Engineering Research Laboratory, under EPA Contract No. 68-03-3069. June 21, 1985b.

Jensen, Daniel. GA Technologies, Inc. Telephone Conversation with Lisa Farrell, GCA Technology Division, Inc. February 4, 1986.

Johanson, J.G., S.J. Yosim, L.G. Kellog, and S. Sudar. Elimination of Hazardous Waste by the Molten Salt Destruction Process. In: Incineration and Treatment of Hazardous Waste, Proceedings fo the Eighth Annual Research Symposium, EPA-600/9-83-003. April 1983.

Johanson, Kenneth. Shirco Infrared Systems, Inc., Dallas, Texas. Telephone Conversations with Lisa Farrell, GCA Technology Division, Inc. February 20, 1986; April 1, 1986; April 3, 1986; April 23, 1986; and May 5, 1986.

Josephson, Julian. Supercritical Fluids. Environmental Science & Technology. October 1982.

Killiley, William. Modar, Inc. Telephone Conversation with Lisa Farrell, GCA Technology Division, Inc. February 25, 1986.

Kolak, Nicholas P., Thomas G. Barton, C.C. Lee, and Edward F. Peduto. Trial Burns - Plasma Arc Technology. In: Proceedings of the U.S. EPA Twelfth Annual Research Symposium on Land Disposal, Remedial Action. Incineration and Treatment of Hazardous Waste. Cincinnati, Ohio. April 21-23, 1986.

Krogh, Charles. CH2M Hill. Memorandum to Katie Biggs (EPA VII), Steve Wilhelm (EPA/VII), and John Kingscott (EPA/HQ). Re: Technical Briefing on the Mobile Incinerator Project. June 7, 1985.

Lee, Anthony. Technical Resources, Inc. Analysis of Technical Information to Support RCRA Rules for Dioxins-containing Waste Streams. Final Draft Report submitted to Paul E. des Rosiers, Chairman, U.S. EPA - Dioxin Advisory Group. July 31, 1985.

Lee, Kenneth W., William R. Schofield, and D. Scott Lewis. Mobile Reactor Destroys Toxic Wastes in "Space". Chemical Engineering. April 2, 1984.

Leslie, R.H. Development of Mobile Onsite Earth Decontaminator, In: Proceedings of the Hazardous '85 Conference, Baltimore, Maryland. May 1985.

Materials Characterization Center (MCC). Nuclear Waste Materials Handbook--Waste From Test Methods. Department of Energy, Washington, D.C. DOE/TIC-11400. 1981.

Marson, L. and S. Unger. Hazardous Material Incinerator Design Criteria, prepared by TRW, Inc. for the U.S. EPA Office of Research and Development, EPA-600/2-79-198. October 1979.

McCormick, Robert. ENSCO, Inc. Telephone Conversation with Lisa Farrell, GCA Technology Division, Inc. May 7, 1986.

McGaughey, J.F., M.L. Meech, D.G. Ackerman, S.V. Kulkarni, and M.A. Cassidy. Radian Corporation. Assessment of Treatment Practices for Proposed Hazardous Waste Listings F020, F021, F022, F023, F026, F027, and F028. Prepared for U.S. EPA under EPA Contract No. 68-02-3148, Work Assignment No. 10. September 1984.

M.M. Dillon, Ltd. Destruction Technologies for Polychlorinated Biphenyls (PCBs). Prepared for Environment Canada, Waste Management Branch. 1983.

Modar, Inc. Brochure: Process Description and Test Results. 1984.

Modell, Michael, Gary G. Gaudet, Glenn T. Hong, Morris Simson, and Klaus
     Biemann. Modar, Inc. Supercritical Water Testing Reveals New Process
     Holds Promise. Solid Waste Management. August 1982.

Modell, Michael, Gary G. Gaudet, Morris Simson, Glenn T. Hong, and Klaus
     Biemann. Modar, Inc. Destruction of Hazardous Waste Using Supercritical
     Water. In: Proceedings of the Eighth Annual Research Symposium on
     Incineration and Treatment of Hazardous Waste, sponsored by U.S. EPA-MERL
     and U.S. EPA-IERL. Ft. Mitchell, Kentucky, March 8-10, 1982.
     EPA-600/9-83-003. September 1983.

Modell, Michael, and Terry B. Thomason. Modar, Inc. Supercritical Water
     Destruction of Dilute Aqueous Wastes. In: Proceedings of the 2nd
     International Symposium on Operating European Centralized Hazardous
     (Chemical) Waste Management Facilities. Odense, Denmark. September 1984.

Oma, K.H., et al. In Situ Vitrification of Transuranic Wastes: Systems
     Evaluation and Applications Assessment. Pacific Northwest Laboratory,
     Richland, Washington. PNL-4800. 1983.

Oma, K.H., R.K. Farnsworth, and C.L. Timmerman. Characterization and
     Treatment of Gaseous Effluents from In Situ Vitrification. In:
     Radioactive Waste Management and the Nuclear Fuel Cycles, Volume 4.
     Hardwood Academic Publishers. 1984.

Poppiti, James. U.S. EPA. Memorandum, Re: RCRA Dioxin Delisting Petition
     for the Mobile Incinerator System. May 23, 1985.

Pyrolysis Systems, Inc. Pyroplasma Waste Management Systems. Product
     Literature.

Pyrotech Systems, Inc. Mobile Waste Processor: MWP-2000-ER. 1985.

Rasmussen, George P. Waste-Tech Services, Inc. Another Option: Onsite
     Fluidized Bed Incineration. Hazardous Materials & Waste Management
     Magazine. January-February 1986.

Rickman, William S., Nadine D. Holder, and Derrell T. Young. GA Technologies,
     Inc. Circulating Bed Incineration of Hazardous Wastes. Chemical
     Engineering Progress. March 1985.

Rollins Environmental Services, Inc. Brochure on Rollins Environmental
     Services and Capabilities. 1985.

Ross, Robert W., II, Frank C. Whitmore, and Richard A. Carnes. Evaluation of
     the U.S. EPA CRF Incinerator as Determined by Hexachlorobenzene
     Incineration. Hazardous Waste, 1(4): 581-597. 1984.

Ross, R.W., II, T.H. Backhouse, R.N. Vogue, J.W. Lee, and L.R. Waterland. Acurex Corporation, Energy & Environmental Division, Combustion Research Facility. Pilot-Scale Incineration Test Burn of TCDD-Contaminated Toluene Stillbottoms from Trichlorophenol Production from the Vertac Chemical Company. Prepared for U.S. EPA, Office of Research and Development, Hazardous Waste Engineering Research Laboratory, under EPA Contract No. 68-03-3267, Work Assignment 0-2. Acurex Technical Report TR-86-100/EE. January 1986.

Roy F. Weston, Inc. and York Research Consultants. Times Beach, Missouri: Field Demonstration of the Destruction of Dioxin in Contaminated Soil Using the J.M. Huber Corporation Advanced Electric Reactor. February 11, 1985.

SCA, Inc., Customer Service Department. Telephone Conversation with Lisa Farrell, GCA Technology Division, Inc. Re: PCBs Incineration Costs. May 9, 1986.

Schofield, William R., Oscar T. Scott, and John P. DeKany. Advanced Waste Treatment Options: The Huber Advanced Electric Reactor and The Rotary Kiln Incinerator. Presented HAZMAT Europa 1985 and HAZMAT Philadelphia 1985.

Shih, C.C., et al. Comparative Cost Analysis and Environmental Assessment for Disposal of Organochloride Wastes. Prepared by TRW, Inc. for the U.S. EPA Office of Research and Development, EPA-600/2-78-190. August 1978.

Shirco Infrared Systems, Inc. Dallas, Texas. Brochure: Shirco Incineration System - Process Description and Component Description. 1985.

Sickels, T.W. ENSCO's Modular Incineration System: An Efficient and Available Destruction Technique for Remedial Action at Hazardous Waste Sites. Preprinted Extended Abstract of Paper Presented Before the Division of Environmental Chemistry, American Chemical Society, 191st National Meeting, New York, New York: Vol 26, No. 1. April 13-18, 1986.

Sieber, F. Modar, Inc. Correspondence with N.F. Surprenant, GCA Technology Division, Inc. May 29, 1986.

Smith, Robert L., David T. Musser, Thomas J. DeGrood. ENRECO, Inc. In Situ Solidification/Fixation of Industrial Wastes. In: Proceedings of the 6th National Conference on Management of Uncontrolled Hazardous Waste Sites, Washington, D.C. November 4-6, 1985.

Spooner, Philip A. Science Applications International Corporation (SAIC). Stabilization/Solidification Alternatives for Remedial Action.

Strachan, D.M., R.P. Turcotte, and B.O. Barnes. MCC-1: A Standard Leach Test for Nuclear Waste Forms. Pacific Northwest Laboratory, Richland, Washington. PNL-SA-8783. 1980.

Swanson, M.L., J. Dollimore, and H.H. Schobert. University of North Dakota,
    Energy Research Center. Supercritical Solvent Extraction. Prepared for
    U.S. Department of Energy. DOE/FE/60181-96. June 1984.

Technical Resources, Inc. Analysis of Technical Information to Support RCRA
    Rules for Dioxin-Containing Waste Streams. Submitted to Paul E. des
    Rosiers, U.S. EPA Office of Research and Development. July 31, 1985.

Thomason, Terry B., and Michael Modell. Modar, Inc. Supercritical Water
    Destruction of Aqueous Wastes. Hazardous Waste, 1(4): 453-467. 1984.

Tittlebaum, Marty E., et al. State-of-the-Art on Stabilization of Hazardous
    Organic Liquid Wastes and Sludges. In: Critical Reviews in
    Environmental Control, 15(2): 179-211. 1985.

U.S. EPA. Office of Research and Development. At-Sea Incineration of
    Herbicide Orange Onboard the M/T Vulcanus. Prepared by TRW Inc., Redondo
    Beach, California. EPA-600/2-78-086. April 1978.

U.S. EPA. Process Design Manual for Sludge Treatment and Disposal,
    EPA-625/1-97-011. September 1979.

U.S. EPA. Fact Sheet on the U.S. EPA Mobile Incineration System (MIS).
    April 1982.

U.S. EPA. Office of Research and Development. At-Sea Incineration of
    PCB-Containing Wastes Onboard the M/T Vulcanus. Prepared by TRW Inc.,
    Redondo Beach, California. EPA-600/7-83-024. April 1983.

U.S. EPA, Region VII. Status Reports on the Mobile Incinerator.
    June-December 1985.

Vick, W.H., S. Denzer, W. Ellis, J. Lambach, and N. Rottunda. Evaluation of
    Physical Stabilization Techniques for Mitigation of Environmental
    Pollution from Dioxin-Contaminated Soils. Interim Report: Summary of
    Progress-To-Date. Submitted to EPA-HWERL by SAIC/JRB Associates. EPA
    Contract No. 68-03-3113, Work Assignment No. 36. June 1985.

Vrable, D.L., and D.R. Engler. GA Technologies, Inc. Transportable
    Circulating Bed Combustor for the Incineration of Hazardous Waste.
    Storage & Disposal. 1985a.

Vrable, D.L., D.R. Engler, and W.S. Rickman. GA Technologies, Inc.
    Application of Transportable Circulating Bed Combustor for Incineration
    of Hazardous Waste. Presented HAZMAT 1985, West Long Beach, California.
    December 1985b.

5. Nonthermal Technologies for Listed Dioxin Wastes

This section reviews nonthermal technologies for treating dioxin wastes. Several of the technologies involve the addition of chemical reagents to degrade or destroy dioxin, e.g. chemical dechlorination, ruthenium tetroxide degradation, and degradation using chloroiodides. Two technologies, ultraviolet (UV) photolysis and gamma ray radiolysis, involve the application of electromagnetic radiation to break down dioxin and other contaminants. There is also a subsection covering biodegradation. The remaining two treatment technologies discussed, solvent extraction and stabilization/fixation, are not destructive technologies, but rather represent pretreatment and temporary measures, respectively, for managing dioxin wastes.

The technologies that are included in this section are not evaluated in the same manner as the thermal technologies. Thermal technologies are primarily evaluated on the basis of six nines DRE, where DRE is a function of the concentration of a contaminant in the exhaust gas from the process. With nonthermal treatment there are generally no exhaust gases of significance. There are, however, other treatment effluents and residues which are major potential sources of dioxin emissions. Since EPA has proposed that these residues must contain less than 1 ppb of CDDs or CDFs in order to be land disposed, this will be the main criterion on which these technologies will be judged.

5.1 CHEMICAL DECHLORINATION

5.1.1 Process Description

Briefly stated, chemical dechlorination processes use specially synthesized chemical reagents to destroy hazardous chlorinated molecules, or detoxify them to form other compounds which can be considered less harmful and environmentally safer than the original hazardous chemical (Dillon, 1982). The basic chemical principle of this dechlorination process involves the gradual, but progressive, substitution of the contaminants' chlorine atoms by other atoms (predominantly hydrogen). This substitution process eventually "deactivates" the previously hazardous chlorinated-contaminant. (It should be noted that dechlorination of halogenated aromatics is not a new principle but rather a common industrial process used for production of phenolics and certain pesticides.)

Many researchers have, over the past years, tested various chemical reagents for use in destroying and/or detoxifying hazardous, chlorinated compounds. This research has evolved into several processes shown to destroy polychlorinated biphenyls (PCBs) and dioxin contaminants. The following discussion provides a review of each of these processes, starting with those dechlorination processes which have been evaluated/designed for PCBs and concluding with more recent studies on dioxin dechlorination from a soil matrix.

One of the first major PCB-contaminated oil dechlorination processes was developed by the Goodyear Tire and Rubber Company (desRosiers, 1983; Weitzman, 1982). This process was intended to remove PCBs from heat transfer fluids using sodium naphthalene as the reagent. The sodium naphthalene reagent is prepared by complexing naphthalene and metallic sodium with the solvent tetrahydrofuran (THF). The reagent mix is reacted with the PCB-contaminated fluid at ambient temperature and at a reagent to chlorine ratio of 50-100:1 (under a nitrogen blanket). Under these conditions, the PCB molecule is stripped of chlorine to form sodium chloride and polyphenyls which, after quenching, are vacuum distilled in order to recover the THF and naphthalene.

The use of the priority pollutant, naphthalene, proved to be a source of concern. Subsequent dechlorination processes were designed to utilize alternate reagents. For example, the Acurex Waste Technologies Corporation (now Acurex Corporation) developed a process (Dillon, 1982; desRosiers, 1983; Weitzman, 1986) which used a sodium-based reagent, prepared from proprietary but nonpriority pollutant constituents (Miille, 1981). The system operates by mixing filtered, PCB contaminated oil with the sodium-based reagent in processing tanks where the chemical reaction occurs. The two streams leaving the reactor are a treated oil containing no PCBs and a sodium hydroxide effluent. The entire PCB destruction process was designed to occur under an inert nitrogen atmosphere, however, Acurex found that this inert nitrogen blanket was not essential (Weitzman, 1986).

The SunOhio PCBX process, approved by USEPA in 1981, is a continuous, closed loop process utilizing a proprietary reagent to strip chlorine atoms from PCB molecules, converting the PCBs to metal chlorides and polyphenyl (polymer) compounds (Dillon, 1982; Jackson, 1981; SunOhio, 1985). PCB-contaminated mineral/bulk oils are first treated to remove moisture and gross contaminants. The PCB-contaminated oil is then mixed with the proprietary reagent and sent to the reactor where PCB destruction occurs. The mixture is then centrifuged, filtered, and vacuum-degassed. Effluent streams include treated oil, and polyphenyl/salt residues. The latter are solidified and then typically sent to a landfill. The entire system is mobile, as it is mounted on 40-foot tractor trailers. SunOhio commercially operates five mobile units.

Only limited information has been found relating to the PPM process; however, more is expected in the near future. From the available information, this mobile process destroys PCB contaminated oil through the aid of a proprietary sodium reagent (M. M. Dillon, 1982; des Rosiers, 1983). The reagent is added to the contaminated oil and left to react. The solid polymer formed by the reaction is filtered out of the oil. While this polymer is reportedly a regulated substance, it has been found to contain no PCBs and can readily be disposed of (M. M. Dillon, 1982). PPM currently has under development a dechlorination process designed to work on soils. However, no information is available on the process (personal conversation with L. Centofanti, PPM).

In 1978, the Franklin Research Institute began studies to develop a chemical reagent that would lead to the cleavage of the carbon halogen bonds inherent with PCBs (Rogers, C., 1983; Klee, A., et al., 1984). Their research identified a chemical reagent which could be synthesized from sodium, polyethylene glycols, and oxygen. This dehalogenation reagent, termed NaPEG, was formulated by mixing molten sodium (60 grams) with 1 liter of polyethylene glycol (PEG) having an average molecular weight of 400. Laboratory studies in 1979 effectively utilized this NaPEG reagent on dielectric fluids containing PCBs, demonstrating the applicability of NaPEG as a dehalogenation reagent.

In 1982, this reagent (generically referred to now as "APEG-alkali polyethylene glycolates") was applied to dioxin-contaminated soils. This research, conducted by the U.S. EPA, Industrial Environmental Research Laboratory in conjunction with Wright State University, was undertaken to establish the effectiveness of these newly-developed APEG reagents. Results clearly indicated that APEG reagents could, under certain laboratory conditions, significantly reduce the levels of TCDD (dioxin) in contaminated soils. The success of these studies has led the U.S. EPA, in cooperation with Galson Research Corporation, to further evaluate the APEG chemical dechlorination process for the destruction/detoxification of dioxins in soil (Peterson, et al., 1985 and 1986). These studies are now in progress.

Scientists in Italy (specifically, at the Institute of Organic Chemistry/University of Torino and the Sea Marconi Technologies Group) have also recently carried out laboratory research using APEGs for the chemical degradation of 2,3,7,8-TCDD. Their process is similar to the Galson Research/USEPA process, but PEGs of much higher molecular weight (1,500 to 6,000 versus 400) are used. The PEGs are then combined with a weak base such as potassium carbonate and an inorganic peroxide such as sodium peroxide to form a clear solution which promotes organic dehalogenation (Tundo, P., et al., 1985). This process, currently referred to as the Sea Marconi's CDP-Process, was first conceived for the decontamination of mineral oils contaminated by PCBs, but has also been shown to destroy 2,3,7,8-TCDD in solvent, in soil or on surfaces as tested at Seveso, Italy (Tumiatti, W., et al., 1986).

5.1.2   Technology Performance Evaluation

The Acurex process, while only applicable to contaminated PCB oils, has been extensively evaluated and is now commercially available via Chemical Waste Management (Weitzman, 1986). Tests by Acurex, during an EPA demonstration in the early 1980's, proved that this technology is effective in treating PCB-contaminated oils containing approximately 1,000 ppm to 10,000 ppm (1%)--reducing the PCB concentration to below detectable limits, about 1 ppm (Weitzman, 1982).

The SunOhio PCBX process has and will continue to be used only on liquid hydrocarbon streams (i.e., oil). The process cannot be used to treat contaminated soils. At present no tests have been performed on 2,3,7,8-TCDD, and it appears that this technology will continue to be used only on PCB contaminated oils or fluids (SunOhio, 1984). This process has reduced PCB contaminated transformer oil from 500 ppm to below detectable limits (1 ppm) in just one pass through the system (Weitzman, 1982). By passing a contaminated oil through the system three times, it is believed that the PCB concentration can be reduced from 3,000 ppm to below 2 ppm [Dillon, 1982; Jackson, 1981].

Performance data regarding the Goodyear process are limited. However, available information indicates that this process is capable of treating oils containing 300 to 500 ppm PCBs down to less than 10 ppm (Weitzman, 1982; Berry, 1981). The contaminated transformer or heat transfer oil is purified to this level within approximately 1 hour at ambient temperature. Like the Acurex/Chemical Waste Management and SunOhio dechlorination processes, the Goodyear process has only been demonstrated to be applicable to treatment of PCB-contaminated oils.

The "APEG-type" processes have been laboratory and, in select situations, field tested on PCBs and 2,3,7,8-TCDD-contaminated soil samples (Klee, A., et al., 1984; Peterson, R., et al., 1985 and 1986; Rogers, C. J., et al., 1985; Rogers, C. J., 1983). The APEG reagents used in these experiments have varied over the several recent years of research from the NaPEG-type, sodium-based reagents used in PCB destruction, to the KPEG-type, potassium-based reagents proven more efficient in 2,3,7,8-TCDD-type destruction.

The results of several of the early 1980 laboratory research and later laboratory/field testing of the APEG chemical dechlorination process are promising. Specifically, in the 1982 research study conducted by the U.S. EPA/Wright State University, actual dioxin-contaminated soils were effectively dechlorinated under certain laboratory conditions (Klee, A., 1984). As shown in Tables 5.1.1 and 5.1.2, dioxin (TCDD) levels in soils were reduced by 8 to 51 percent depending on the specific APEG reagent and number of days after application that the dioxin levels were measured. A multiple application experiment of the K-400 (potassium-based reagent and polyethylene glycol of average molecular weight of 400) reagent showed that an increase of from 16 to 56 percent and 25 to 68 percent could be realized by repeat application versus a single application to the contaminated soil.

Later laboratory research, in 1985, by the U.S. EPA/Galson Research Corporation (using 1,2,3,4-tetrachlorodibenzo-p-dioxin) demonstrated that chlorinated dioxin levels in soil may be further chemically reduced by applying APEG-type reagents (Peterson, R.L., et al., 1985). In situ and slurry testing, using potassium hydroxide/polyethylene glycol 400/dimethyl sulfoxide (KOH/PEG/DMSO) and potassium hydroxide/2-(2-methoxy ethoxy ethanol)/dimethyl sulfoxide (KOH/MEE/DMSO) reagents, on contaminated soils containing an initial concentration of 2000 ppb was quite favorable, as summarized in Tables 5.1.3 and 5.1.4. Several key features uncovered during the experiments are as follows:

- Temperature increases from 20 to 70°C during the in situ process indicated a dramatic improvement in reaction efficiency, i.e., from 50 to 90 percent increase.

- No difference between reagent formulations was noted at 70°C during in situ testing.

- Dilution of the reagent with water (to provide more contact, followed by evaporation of the water to encourage reaction) was not effective in reducing the amount of reagent required during the in situ processing.

- A removal efficiency of 99.95% TCDD (from 2000 ppb to 1 ppb) was realized after 12 hours at 70°C during the slurry processing.

TABLE 5.1.1.   SUMMARY OF DATA SHOWING PERCENT REMOVAL OF TCDD
FROM CONTAMINATED SOILS USING APEG DECHLORINATION PROCESS
(Klee, A. et.al., 1984).

Days after application	Timberline[a]		Denny[b]	
	K-400	K-120	K-400	K-120
7 days	45%	46%	nm[c]	51%
28 days	35%	38%	12%	5%

[a]Initial TCDD content equalled 277 28 ppb.

[b]Initial TCDD content equalled 330 33 ppb.

[c]nm = not measured

TABLE 5.1.2.   SUMMARY OF DATA SHOWING PERCENT REMOVAL OF TCDD
FROM CONTAMINATED SOIL AT DENNY FARM
(Klee, A., et. al., 1984)

Days after application	Denny Farm Soil[a]	
	K-400[b]	KM-350
1 day	8%	15%
7 days	19%	27%
14 days	16%	36%
21 days	25%	42%
28 days	22%	43%

[a]   The anomalies in the apparent decrease of the TCDD level of K-400 treated
sample at day 14 was found not to be statistically significant.

[b]   K-400 reagent used in these experiments (vs. those shown in previous
Table 5.1.1) was prepared from KOH pellets instead of a 66% aqueous KOH
solution.

TABLE 5.1.3.   SUMMARY OF RESULTS OF IN-SITU PROCESSING (PETERSON, R. L., et. al., 1985) - ALL SOILS INITIALLY AT 2000 ppb.

Reagent	Wt% in soil	Temp, (°C)	Time, (days)	Final TCDD (avg) Concentration (ppb)
1:1:1 KOH/PEG/DMSO	20	20	7	980
1:1:1 KOH/PEG/DMSO	20	70	7	<1
1:1:1 KOH/PEG/DMSO	20	70	1	5.3
2:2:2:1 KOH/MEE/DMSO/WATER	20	70	1	3.3
2:2:2:1 KOH/MEE/DMSO/WATER	20	70	2	2.8
2:2:2:1 KOH/MEE/DMSO/WATER	20	70	4	2.1
2:2:2:1 KOH/MEE/DMSO/WATER	20	70	7	1.2
2:2:2:6 KOH/MEE/DMSO/WATER	20	70	7	2.1
2:2:2:30 KOH/MEE/DMSO/WATER	50	70	7	18
2:2:2:30 KOH/MEE/DMSO/WATER	20	70	7	50
BLANKS - ALL				<1

TABLE 5.1.4.   RESULTS OF SLURRY PROCESSING (PETERSON, R. L., et. al., 1985).

Reagent	Temp, °C	Reaction Time, hrs	Final TCDD Concentration, ppb
1:1:1 KOH/PEG/DMSO	180-260	4	<1
1:1:1 KOH/PEG/DMSO	180	2	<1
1:1:1 KOH/MEE/DMSO	150	2	<1
1:1:1 KOH/MEE/DMSO	70	2	<1
1:1:1 KOH/MEE/DMSO	70	0.5	15
1:1:1 KOH/MEE/DMSO	25	2	36

Blanks - all <1 ppb TCDD

Spikes - % recovery in soil - 0.1-5.9

Additionally, it should be noted that during the summer of 1985, APEG-type reagents were tested by the U.S. EPA at the Shenandoah Stables dioxin-contaminated site to evaluate the dechlorination potential of these reagents on 2,3,7,8-tetrachlorodibenzo-p-dioxin (2,3,7,8-TCDD) under <u>field conditions</u> (Rogers, C. J., 1985). Results of these tests were not as promising as in the past using APEGs in the laboratory. Specifically, the APEG reagents were deactivated due to the fact that APEG is moisture sensitive. The soil moisture at Shenandoah was determined to be on the order of 18 to 21 percent by weight. These results, while not favorable, did point out that APEGs are extremely hygroscopic and that contact with moisture will eventually result in the deactivation of the APEG reagent.

Finally, results of the use of the Sea Marconi CDP-process are presented in Figure 5.1.1 and Table 5.1.5. Figure 5.1.1 shows that, at least in the beginning, the disappearance of 2,3,7,8-TCDD from the reaction mixture is linear with respect to time (Tundo, P., et al., 1985). The figure also indicates that mixtures containing higher weight PEGs promote a much more rapid decomposition of TCDD than mixtures containing lower weight PEGs. When a PEG with a molecular weight of 6,000 is used (square data points), greater than 99 percent decomposition of TCDD occurs in 30 minutes, while the use of a PEG with a molecular weight of 1,500 (circular data points) requires over two hours for an equivalent level of degradation. However, the reaction rate is also a function of temperature, and the reaction using PEG 6000 was carried out at 85°C versus 50°C for the PEG 1500 reaction. Therefore, based on the data presented in Figure 5.1.1, it is difficult to assess the full effects of higher molecular weight PEGs.

The data in Table 5.1.5 show the effect of temperature and different mixtures of reagents on the decomposition of TCDD. The first set of data represents the same conditions (PEG 6000 at 85°C) as those used to generate the square data points in Figure 5.1.1. As already mentioned, these conditions result in rapid decomposition of TCDD. The second set of data points was generated using a smaller quantity of PEG 6000 and adding a butyl ether compound to the reaction mixture. With this combination of reagents, greater than 99.9 percent decomposition of TCDD occurred in 30 minutes. For the third set of data, the reaction temperature was only 20°C, and the decomposition of TCDD was much slower than for all of the other uses. Only 50 percent decomposition occurred in 192 hours.

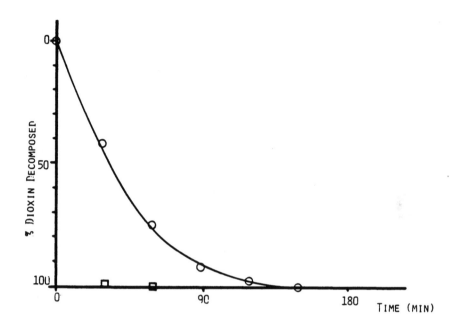

T = 50°C; 2.0 ml of n-decane containing 5 µg of dioxin,
0.9 g PEG 1500, 0.15 g K_2CO_3, 0.10 g of an ether compound
and 0.10 g Na_2O_2

T = 85°C; 2.0 ml of n-decane containing 5 µg of dioxin,
2.06 g of PEG 6000, 0.53 g K_2CO_3, and 0.37 g Na_2O_2

Figure 5.1.1.   Degradation of 2,3,7,8-TCDD using the
CDP-Process (Tundo, P., et al., 1985).

TABLE 5.1.5.   DEGRADATION OF 2,3,7,8-TCDD UNDER DIFFERENT CONDITIONS USING THE CDP-PROCESS (Tundo, P. et al, 1985)

REAGENTS (g)		TEMPERATURE (°C)	TIME (h)	DECOMPOSITION (%)
PEG 6000	(2.1)	85	0.5	99.4
K_2CO_3	(0.5)		1.0	99.6
Na_2O_2	(0.4)		1.5	99.75
			2.0	>99.9
PEG 6000	(1.3)	85	0.5	>99.9
K_2CO_3	(0.5)			
Na_2O_2	(0.2)			
$BuO(CH_2CH_2O)_2H$	(0.2)			
PEG 6000	(1.8) [a]	20	72	30
K_2CO_3	(0.4)		192	50
Na_2O_2	(0.2)			

[a] without n-decane: after homogenization at 80°C the reaction was solidified by cooling and kept at 20°C;

The data indicate that the CDP process is effective in decomposing TCDD under laboratory scale conditions. However, researchers have questioned the effectiveness of high molecular weight PEGs under field conditions. Their skepticism concerns the ability of the high molecular weight PEG to penetrate and react with TCDD in soil micropores (Technical Resources, Inc. 1985). These concerns will have to be addressed by further research.

5.1.3   Cost of Treatment

At this time, costs are very well established for the decontamination of PCB contaminated oils. These costs are dependent on several variables:

- concentration of pollutant;

- quantity and characteristics of the material to be treated;

- reagent costs; and

- the resale value of the treated material.

The cost of treating PCB-contaminated oil using the SunOhio PCBX process is reportedly about $3/gallon of contaminated oil in bulk. However, the cost for onsite treatment of transformers may be considerably more depending on the site specific situation (Marilee Fisher, SunOhio).

Based upon the APEG laboratory (and select field) research that has been conducted over the past several years, a preliminary economic evaluation of this dechlorination process has been attempted (Peterson, R. L., et al., 1985). Specifically, Galson Research Corporation, in conjunction with the U.S. EPA-HWERL, has roughly estimated the costs for APEG dechlorination using two hypothetical field scenarios. These costs, as shown in Table 5.1.6, indicate that for the in situ process (operating on a 1 acre-3 feet deep contaminated area) vs. the slurry process (with excavation and 3 reactor systems operating) there is approximately a $205/ton difference. This difference comes from the fact that in the slurry APEG process, reagent recovery is a component which accounts for recovery of approximately 65 percent of the total cost of the in situ process. Another source has restated these above costs on the basis of $/acre/cm indicating that the slurry process costs are approximately $1,000/acre/1 cm penetration (Technical Resources Inc., 1985).

TABLE 5.1.6.   PRELIMINARY ECONOMIC ANALYSIS OF
IN SITU AND SLURRY PROCESSES
(Peterson, R.L., et al., 1985)

Cost item	Cost, $/ton soil	
	In situ	Slurry
Capital recovery	31	17
Setup and operation	65	54
Reagent	200	20
Total costs	296	91

5.1.4 Process Status

As stated previously, the chemical dechlorination processes developed by Acurex (Chemical Waste Management), SunOhio and Goodyear are exclusively for the treatment of liquid PCB-contaminated oils.   In fact, it was noted that the sodium-based reagent process developed by Acurex Corporation should never be used in the field on soils due to its explosive nature (Weitzman, 1986).

On the other hand, research using variations of APEG reagents has been and will continue. Specifically, Galson Research Corporation is continuing laboratory and field testing of the treatment of dioxin contaminated soils (Peterson, R. L., et al., 1986). This research testing was conducted within 55-gal. mixing reactors located at Pine Bluff, Arkansas (Peterson, R.L., 1986a). In these reactors, the slurry process was used on soils initially containing 100-150 ppb of dioxin and allowed to react for approximately 3 hours at 100-150°C. Results from this research showed a removal efficiency of 99 percent.

It is currently envisioned that this research will be continued using the Electric Power Research Institute's 100-gal reactor placed on a Galson trailer (Peterson, R.L., 1986a). These tests are to be carried out in parallel with an incinerator receiving the same feed stock of contaminated soil.

Due to inability of the in situ process to reduce dioxin levels in soil to below 1 ppb, it is expected that this variation of the technology will be applied instead to PCB contaminated soils where residual levels of contaminant need not be so low.

5.2 UV PHOTOLYSIS

5.2.1 Process Description

Ultraviolet (UV) radiation is electromagnetic radiation having a wavelength shorter than visible light, but longer than x-ray radiation. The energy content of light increases as the wavelength decreases. For wave lengths in the UV region the energy is sufficient to break chemical bonds and bring about rearrangement or dislocation of molecular structures. The energy

corresponding to the absorption of a quantum (photon) of light is 95 kilo
calories per gram-mole for UV light with a wave length of 3,900 angstroms and
is 142 kilocalories per gram-mole for a wave length of 2,000 angstroms.

Table 5.2.1 lists the dissociation energies for many common chemical
bonds, along with the wavelength corresponding to the energy at which UV
photons will cause dissociation. As can be seen from the data in Table 5.2.1,
bond dissociation energies range from a low of 47 kcal/gmole for the peroxide
bond to a high of 226 kcal/gmole for the nitrogen triple bond. Of particular
interest in the case of dioxins is the C-Cl bond, with a dissociation energy
of 81 kcal/gmole, corresponding to an optimum UV wavelength of 353 nm. For
reference purposes, this can be compared to the violet end of the visible
spectrum with a wavelength of about 420 nm. Thus, the UV radiation of
interest is in the electromagnetic spectrum close to visible light. This fact
is important because it means that sunlight, which radiates strongly in the
near visible wavelengths, might be a good source of UV photons which are
capable of degrading many molecules.

It is not surprising then, that the use of sunlight to degrade certain
toxic molecules has been noted by several researchers (des Rosiers,
P.E., 1983; Zepp, R.G., 1977, Esposito, M.P., 1980; Crosby, D.G., 1971). In
the case of 2,3,7,8-TCDD and other related compounds, the apparent mechanism
is that a terminal C-Cl bond is broken by UV radiation, thus "dechlorinating
the molecule" and converting it into less toxic compounds. (Note that this
reaction mechanism is very similar to that of chemical dechlorination; i.e., a
gradual and progressive substitution of the chlorine atoms.)

Efficient degradation appears to require the presence of a hydrogen
donor, because while UV can cause the cleavage of the C-Cl bond, recombination
can take place. However, if a hydrogen donor is present, it will also react
and replace the chlorine on the molecule. For example, several researchers
have noted that pure 2,3,7,8-TCDD and other chlorinated compounds degrade
slowly or not at all when placed on inorganic substrates, but when suitable
hydrogen donors are present, degradation in sunlight can be rapid (Crosby,
1978).

UV has been commercially used to kill micro-organisms such as bacteria,
protozoa, viruses, molds, yeasts, fungi, and algae. Applications include
process and drinking water disinfection and sterilization, pretreatment prior
to reverse osmosis, and general algae and slime control.

TABLE 5.2.1.   DISSOCIATION ENERGIES FOR SOME CHEMICAL BONDS

Bond	Dissociation energy (kcal/gmol)	Wavelength to break bond (nm)
C-C	82.6	346.1
C=C	145.8	196.1
C≡C	199.6	143.2
C-Cl	81.0	353.0
C-F	116.0	246.5
C-H	98.7	289.7
C-N	72.8	392.7
C=N	147.0	194.5
C≡N	212.6	134.5
C-O	85.0	334.5
C=O (aldehydes)	176.0	162.4
C=O (ketones)	179.0	159.7
C-S	65.0	439.9
C=S	166.0	172.2
Hydrogen		
H-H	104.2	274.4
Nitrogen		
N-N	52.0	540.8
N=N	60.0	476.5
N≡N	226.0	126.6
N-H (NH)	85.0	336.4
N-H (NH$_3$)	102.0	280.3
N-O	48.0	595.6
N=O	162.0	176.5
Oxygen		
O-O (O$_2$)	119.1	240.1
-O-O-	47.0	608.3
O-H (water)	117.5	243.3
Sulfur		
S-H	83.	344.5
S-N	115.	248.6
S-O	119.	240.3

Source: Legan, R.W.   1982.

Recently, UV photolysis has been viewed as a potential large-scale commercial mechanism to degrade toxic wastes. In attempting to obtain a simple, inexpensive, and effective soil detoxification method, the University of Rome evaluated the use of various cationic, anionic, and nonionic surfactants to solubilize 2,3,7,8-TCDD in an aqueous solution prior to photodegradation with sunlight or artificial UV light (Botre, C., 1978). Of the four surfactants, 1-hexadecyclpyridium chloride or cetylpyridium chloride (CPC) was found to be the most effective solubilizing agent, as well as having the ability to enhance the subsequent photochemical degradation of 2,3,7,8-TCDD. Other solvents examined included sodium dodecyl sulfate (SDS), polyoxyethylene sorbitan monoleate (Teewn 80) and methanol.

In 1975, Velsicol Chemical Corporation (Chicago) experimented with removing 2,3,7,8-TCDD contamination from stockpiles of "Agent Orange", a defoliant used in Viet Nam (Crosby, D.G., 1978; des Rosiers, P.E., 1983). The 2,3,7,8-TCDD molecule was extracted by using n-heptane as a solvent, and exposing the solution to UV photolysis at 300-320 nm wavelength. The process resulted in a reduction from 1,900 ppb of 2,3,7,8-TCDD in the stockpiles to less than 50 ppb in end products (Zepp, R.G., 1977; Esposito, M.P., 1980). However, the process was not considered practical, and soon was discontinued.

Another UV degradation process that was developed in the early 1980s by the Atlantic Research Corporation was named Light Activated Reduction of Chemicals (LARC). This process involves bubbling hydrogen into an aqueous solution containing chlorinated hydrocarbons and then irradiating the solution with ultraviolet light to declorinate the contaminants. Work on this process was stopped several years ago for economic reasons (Kitchens, 1986).

More recently, three UV based processes have been described in the literature which may be viable for large-scale degradation of 2,3,7,8-TCDD, as well as other toxic chlorinated hydrocarbons. There are:

- the Syntex - IT Enviroscience process which involves UV photolysis preceded by solvent extraction,

- UV photolysis in combination with ozonation, and

- UV photolysis preceded by thermal desorption.

Section 5.2.2 contains a discussion of the performance of these three processes.

5.2.2   Technology Performance Evaluation

Syntex-IT Enviroscience--

     A commercial-scale UV degradation process was developed cooperatively by
Syntex Agribusiness and IT Enviroscience, and used by Syntex on 4,300 gallons
of still bottoms composed of roughly equal concentrations (50-55%) of
trichlorophenols and ethylene glycol derivatives (45-50%) and containing
approximately 340 ppm 2,3,7,8-TCDD. The first step in this process involved
neutralization and extraction of the 2,3,7,8-TCDD from the still bottoms using
sulfuric acid-n-hexane-caustic soda. Multiple (up to eight) extractions were
used to remove as much 2,3,7,8-TCDD as possible. The hexane 2,3,7,8-TCDD
mixture was then placed in a UV batch reactor. Isopropyl alcohol was added as
a hydrogen donor. Each batch was then exposed to UV irradiation for about
27 hours under conditions of turbulent mixing to insure exposure of the
2,3,7,8-TCDD to the UV radiation. Destruction efficiencies of greater than
98.7 percent of the original 2,3,7,8-TCDD were achieved (des Rosiers, P.E.,
1983; Sawyer, C.J., 1982; Exner, J.H., 1982).

     As shown in Figure 5.2.1, the photodecomposition rate of TCDD is
pseudo-first-order over a concentration range of five orders of magnitude.   In
addition, the rate of decomposition was found to be temperature independent,
and the reaction products were less toxic than dioxin (Exner, J.H., 1982).
The only problem with the extraction/UV photolysis process was that four
process waste streams were generated and were still contaminated with
2,3,7,8-TCDD as well as less toxic, but still hazardous materials.

     These process residuals are listed in Table 5.2.2. They are a result
both of the inefficiency of the extraction process and the need to remove all
dioxin from process equipment. Even though a majority of the dioxin was
extracted from the still bottoms, 500 gallons still remained containing 35 ppm
of dioxin. In addition, 5000 gallons of waste were generated by rinsing
equipment that was used for extraction and photolysis. Even though this waste
stream only contained 20 ppb of dioxin, 1 ppb is the level which must be
attained to allow non-hazardous disposal. Consequently, each of these waste
streams was placed in a RCRA interim status storage facility to await final
disposal or destruction.

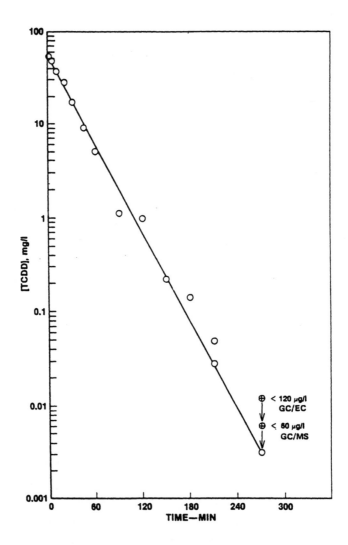

Figure 5.2.1. Rate of dioxin disappearance via UV irradiation of hexane extract of dioxin-contaminated still bottoms (Exner, J.H., 1982).

TABLE 5.2.2.   ESTIMATED VOLUMES AND CONCENTRATIONS OF 2,3,7,8-TCDD
              PRODUCED BY THE SYNTEX-IT PHOTOLYIC PROCESS

	Volume (gallon)	2,3,7,8-TCDD (ppb)
Treated waste	4,000	200
Aqueous (salt) waste	20,000	2
Hexane still bottoms	500	35,000
Equipment solvent rinses	5,000	20

Source: des Rosiers, P.E.   1983.

Ultraviolet (UV) Ozonolysis--

In 1979, it was shown (see Figure 5.2.2) by the California Analytical Laboratories and the Carborundum Company that ultraviolet activated ozone could successfully degrade 2,3,7,8-TCDD from the 1 ppb levels in solution to less than 0.4 ppb (Edwards, B.H., 1983). The procedure utilized to produce the results shown in Figure 5.2.2 consisted of bubbling ozone gas through the TCDD solution, which was then passed by UV lamps. The UV radiation lamps not only degraded the 2,3,7,8,-TCDD directly, but integrated with ozone to enhance the oxidation of the 2,3,7,8-TCDD. No information was available regarding the waste products that were generated from this process.

UV ozonolysis has also been tested extensively in degrading PCBs down to levels of 1 ppb. An "ULTROX" pilot plant at a General Electric (GE) plant in Hudson Falls, New York, and another smaller installation at the Iowa Ammunition Plant, Burlington, Iowa, have proven the technical feasibility of this process on PCBs (Arisman, R.K., 1980; Edwards, B.H., 1983; Swarzgn, E.M., 1982). Both of these plants mixed wastewater containing PCBs with ozone, then exposed the mixture to UV radiation in a mixing tank. Figure 5.2.3 shows a schematic of the pilot plant set up by GE to demonstrate the ULTROX UV/ozone system for PCBs.

Another UV ozonolysis process is called the "Oxyphoton" process. The process was reportedly capable of destroying a wide variety of toxic or organic compounds including PCBs, chlorinated dioxins, DDT, and many types of halogenated aliphatic and aromatic compounds (Worne, 1984). The process is carried out in stainless steel reactors and is capable of treating 60 to 1,800 gallons per hour of waste fluids. Liquid waste containing a proprietary catalyst is spray-atomized and premixed under pressure with oxygen containing 1 to 2 percent ozone prior to passage through the high intensity ultraviolet (UV) light.

One advantage to this vapor phase reaction process over the conventional liquid phase UV light processes is the rapid disintegration of the waste. Reaction rates are generally reported in the millisecond range. Presently, the oxyphoton process is on the "back burner", possibly because of unfavorable economics. No research efforts have directly involved 2,3,7,8-TCDD (Worne, 1984).

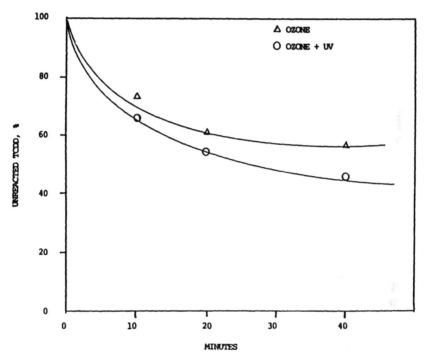

Figure 5.2.2.   Removal of TCDD versus time using
Ozonation/UV irradiation process.
(5 mg Ozone/liter and at a pH of 7.0)
(Edwards, B. H., 1983).

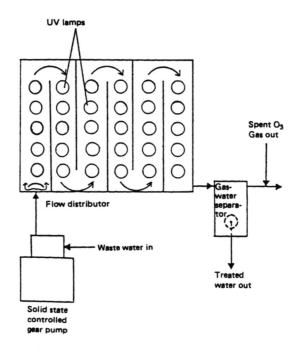

UV lamps

Spent O_3
Gas out

Gas-
water
separa-
tor

Flow distributor

Waste water in

Treated
water out

Solid state
controlled
gear pump

Figure 5.2.3.   Schematic of top view of ULTROX pilot
plant by General Electric (Ozone sparging
system omitted) (Edwards, B. H., 1983).

Thermal Desorption/UV Photolysis--

Based partially upon IT Corporation's past experience in the use of UV photolysis for the chemical degradation of 2,3,7,8-TCDD-contaminated materials (as discussed above), recent laboratory and onsite pilot scale experiments have been undertaken to confirm or deny the applicability of an extension of the UV photolysis process (Technical Resources Inc., 1985; Helsel, R., 1986). Specifically, a system comprised of a thermal desorber, followed by a solvent based absorption/scrubbing system and a UV photolytic unit (as shown in Figure 5.2.4), was tested on contaminated soil containing Herbicide Orange. In the laboratory tests, key process design and operating parameters were established using a C_{12} aliphatic hydrocarbon soltrol 170 (a product of Phillips Petroleum) for absorption/scrubbing. These tests indicated that a TCDD removal efficiency of greater than 99 percent was achieved across both scrubber stages.

Following these tests, a pilot-scale field demonstration system was tested on soils during May-June 1985 at the Naval Construction Battalion Center (NCBC) in Gulfport, MS (Helsel, R., 1986). During these tests, a series of five desorption and three photolysis runs were made. The untreated soil consisted of sand, shells, cement stabilized soil, and traces of asphalt; a composite was prepared, which was considered representative of the various portions of the contaminated site. The composite soil was air dried; crushed to less than 1/4 inch (to enable passage through the desorber feed mechanism); and thoroughly blended with TCDD. Analysis of five separate drums indicated approximately 250 ppb 2,3,7,8-TCDD with good uniformity. The total quantity of soil processed during each run, which lasted from 6 to 12 hours, ranged from 200 to 500 pounds. Samples of treated soil and scrubber solvent before and during photolysis were taken and analyzed for 2,3,7,8-TCDD within 3 days to provide information for adjusting test conditions, as necessary, for succeeding runs.

Results of the pilot-scale field tests indicate that the goal of less than 1 ppb residual 2,3,7,8-TCDD in the soil was achieved for all desorption runs, with desorber operating temperatures ranging from 460° to 500°C and feedrates ranging from 30 to 100 pounds/hour. The photolysis system demonstrated that 99 percent reduction of the 2,3,7,8-TCDD to less than 1 ppb could be achieved in 6 hours, with a corresponding 85 to 95 percent reduction of 2,4,5-T and 2,4-D, respectively.

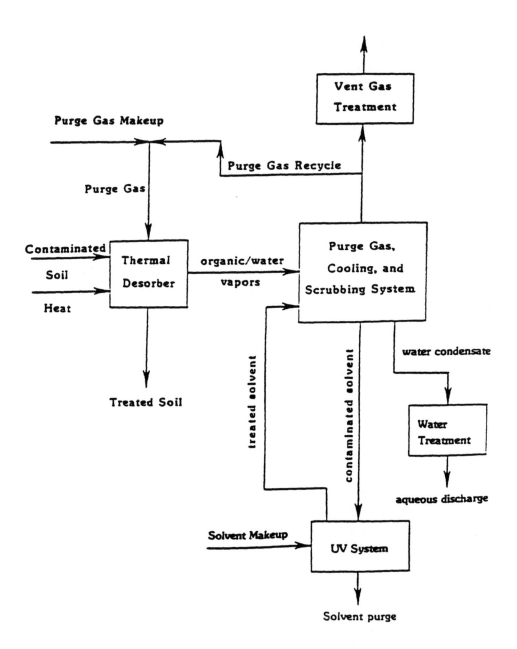

Figure 5.2.4.   Thermal desorption, solvent absorption/scrubbing,
UV photolysis process schematic (des Rosiers,
P. E., 1985).

5.2.3 Cost of Treatment

Several estimates of the cost to treat the 4300 gallons of 2,3,7,8-TCDD contaminated still bottoms using the Syntex-IT process have been presented in the literature. These estimates have ranged from $500,000 (Waste Age, 1980) to over one million dollars (Chemical Engineering, 1980).

The UV ozonation process has been proven effective for aqueous solutions of PCBs, and is reported to be competitive with activated charcoal for cleaning up contaminated wastewater (Edwards, B.H., 1983; Swarzgn, E.M., 1982). Specific capital cost estimates for treating PCBs, suggest that approximately $300,000 is required for a 150,000 gpd wastewater treatment plant, with reduction of PCBs from 50 ppb to 1 ppb (Arisman, R.K., 1980). Table 5.2.3 identifies design specifications and cost data for the ULTROX treatment process. However, as with the direct UV process, it is proven only for liquids and no work on 2,3,7,8-TCDD has been reported.

Finally, precise costs associated with the thermal desorption/UV photolysis system have not yet been developed. It is estimated, however, that the cost would range from $250 to $1,250/ton of soil treated (Technical Resources, Inc. 1985).

5.2.4 Process Status

Presently, little development data and/or progress have been reported on those processes designed primarily for PCBs. However, it has been indicated that further field and demonstration testing of thermal desorption/UV photolysis on Johnston Island in the Pacific was reportedly scheduled for early 1986 (Helsel, R., 1986). During these tests, it is anticipated that a series of seven desorption tests on contaminated coral-derived soil will occur. Following these tests, it is expected that an engineering/economic analysis of the system will be performed.

TABLE 5.2.3.   DESIGN SPECIFICATIONS, CAPITAL, AND O&M COSTS FOR
40,000 AND 150,000 GPD ULTROX TREATMENT PLANTS
(50 ppm PCB feed–1 ppm PCB effluent)

DESIGN SPECIFICATIONS

	40,000 GPD (151,400 LPD) Automated System	150,000 GPD (567,750 LPD) Automated System
Reactor		
Dimension, Meters (LxWxH)	2.5 x 4.9 x 1.5	4.3 x 8.6 x 1.5
Wet Volume, Liters	14,951	56,018
UV Lamps; Number 65 W	378	1,179
Total Power, kw	25	80
Ozone Generator		
Dimensions, Meters (LxWxD)	1.7 x 1.8 x 1.2	2.5 x 1.8 x 3.1
kg Ozone/day	7.7	28.6
Total Energy required (kwh/day)	768	2544

BUDGETARY EQUIPMENT PRICES

	40,000 GPD	150,000 GPD
Reactor	$94,500	$225,000
Generator	30,000	75,000
TOTAL	$124,500	TOTAL $300,000

O & M Costs/Day

Ozone Generator Power	$4.25	$15.60
UV Lamp Power	15.00	48.00
Maintenance (Lamp Replacement)	27.00	84.20
Equipment Amortization (10 Yrs @ 10%)	41.90	97.90
Monitoring Labor	85.71	85.71
TOTAL/DAY	$173.86	$331.41
Cost per 3785 Liters		
(with monitoring labor)	$4.35	$2.21
(without monitoring labor)	$2.20	$1.64

Source: Arisman, R.K. and Musick, R.C., 1980.

5.3   SOLVENT EXTRACTION

5.3.1   Process Description

Solvent extraction, as discussed here, involves contacting contaminated soil with solvents which preferentially desorb the contaminant(s) molecules from the soil matrix. As such, solvent extraction is not a complete treatment technology but a pretreatment step in an ultimate treatment process train.

There are several requirements for the successful application of solvent extraction to dioxin contaminated materials (soils). First, a favorable equilibrium is needed to provide a considerable transfer of the dioxin molecules from the solid phase to the liquid phase . Second, it is important that the rates at which dioxin molecules transfer into the solvent be fast enough so that the overall process occurs in a reasonable length of time. Finally, to achieve high removal efficiencies, the amount of pollutant removed must be proportional to the amount on the soil resulting in a geometric reduction in the concentration of dioxin molecules (Weitzman, 1984).

These requirements suggest that a desirable solvent would have the following properties (Weitzman, 1984; Firestone, 1984):

- the ability to reduce the concentration of the original dioxin molecules in the soil to an acceptable level;

- nonflammability for safety in the field;

- a low latent heat of vaporization and a low boiling point for ease of recovery;

- a low toxicity;

- a low cost;

- commercial availability; and

- compatibility with other treatment schemes, such as photolysis, incineration, or biodegradation following extraction.

Usually more than one wash will be needed, leading to the use of multiple batch processes or continuous countercurrent processes. Soil preprocessing, contacting devices, and solvent recycling are likely features of any solvent extraction unit. Preprocessing will probably involve the particle size

reduction of soils to enhance optimum solvent/soil contact. Solvent recycling
allows reuse of expensive solvents and lowers concerns about disposal of
contaminant-containing solvents. Distillation or vacuum stripping are the
usual methods for cleaning solvents. In either case, the result is a
concentrated volume of contaminant for eventual treatment or disposal
(Weitzman, 1984; Firestone, 1984).

EPA has developed a mobile soils washing system (MSWS) process for
extracting dioxin and other contaminants from soil. The EPA-developed MSWS
contains two basic components, as summarized below from IT Corporation, 1985
and as shown in Figure 5.3.1. These components are a Drum Screen Scrubber and
a Counter-Current Chemical Extractor. The Drum Screen unit automatically
loads previously excavated soil (particle sizes less than 1 inch) into the
system where it passes through high pressure streams of extractant solution
and a "soaking zone". The high pressure streams are designed to wash sands
and stones and to separate fines for further, high energy extraction. Sands
and stones are discharged from the Drum Screen and the fines are pumped
continuously into the Counter-Current Extractor, which consists of four
high-shear mixing chambers. As the fine soil (less than 2 mm) leaves each
chamber, it is separated from its solvent carrier before it enters the next
chamber. The design capacity of the MSWS is 18 cubic yards of soil per hour.

5.3.2 Technology Performance Evaluation

Solvent extraction of chemical substances from soil has been commonly
used in the mining industry and has been demonstrated for extraction of
bitumen from tar sands (Cotter, 1981). Currently, the only full-scale process
that has attempted to use solvent extraction for dioxin molecules dealt with a
contaminated slurry. In this instance, the dioxin molecule (2,3,7,8-TCDD) was
extracted from distillation still bottoms at the Syntex Agribusiness facility
in Verona, Missouri. IT Enviroscience was contracted by Syntex to develop a
safe and effective method for removing approximately 7 kg of dioxin from about
4600 gal of waste (Exner, J.H., et al 1982). The treatment process designed
by IT involved the separation (extraction) of dioxin using a solvent, followed
by the photolytic dissociation of the carbon-halogen bond (see Section 5.2).
The solvent extraction phase of this project, as briefly described below,
involved several laboratory, miniplant, and scale-up operations.

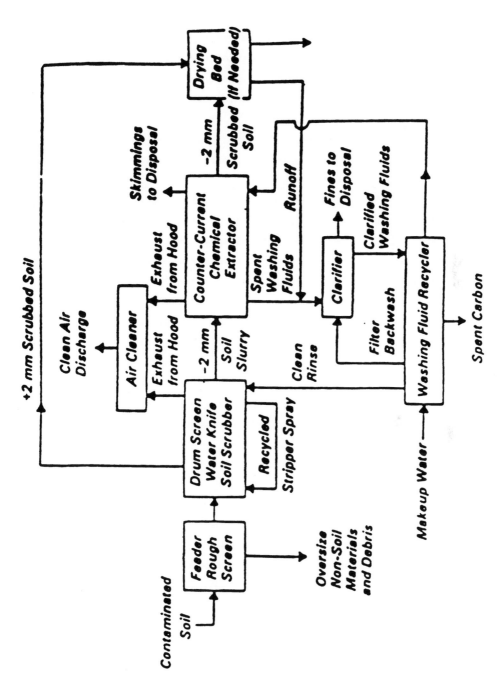

Figure 5.3.1. Process flow scheme for EPA-developed Mobile Soils Washing System (IT Corporation, 1985).

Specifically, IT Enviroscience performed tests on hexane, tetrachloro-
ethylene and o-xylene to determine which solvent would best remove the dioxin
molecule and, once removed, would allow the dioxin molecule to be effectively
degraded via the photolytic step. Their results showed that hexane extraction
of the subject wastes performed better overall than the other two solvents.
Based upon these results, a large-scale reactor vessel was designed and
constructed. In 1980, this reaction vessel processed several 160-gal batches
of the dioxin-containing waste resulting in a reduction of 2,3,7,8-TCDD-
concentrations of from 340 to 0.2 ppm via six hexane extractions.

IT Corporation, under the auspices of EPA, has also prepared additional
laboratory experiments to assess the suitability of the EPA Mobile Soils
Washing System (MSWS) for use in extracting dioxin from contaminated soils (IT
Corporation, 1985). The MSWS was designed to use water, or water with
non-toxic and/or biodegradable additives, as an extractant solution.
Non-hazardous additives are required because some residual solution will
always remain with the discharged soil. Because of this requirement, various
additives, such as surfactants and fuel oil, were evaluated in the laboratory
for the removal of dioxin from soil. Although laboratory results indicated
that 60% to over 90% of the 2,3,7,8-TCDD could be removed by the Soils Washing
System, in most cases (soils initially containing over 100 ppb of dioxin) the
washed soil would still contain residual dioxin in excess of the 1 ppb
guideline for decontamination. Similarly, while other experiments using Freon
and Freon-methanol combinations proved promising, the target residual dioxin
levels could not be achieved under the test conditions. It was concluded that
the major obstacle to removing dioxin from the soils was that dioxin binds
strongly to small soil particles. The soils on which the MSWS was tested were
from the Denney Farm in Missouri. These soils contained a high percentage of
extremely fine materials (33% less than 5 microns, 26% less than 1 micron).
For materials with larger grain size, such as sands and gravels, the process
may be viable (IT Corporation, 1985).

In other laboratory experiments, both aqueous and organic solvents have
been tested on 2,3,7,8-TCDD-contaminated soil. In 1972, Kearney used a 1:1
hexane: acetone solvent solution on 2,3,7,8-TCDD (labeled with carbon
isotopes) in loamy sand and silty clay-loam soils. Electron-capture gas

chromotography revealed that approximately 85 percent of the applied TCDD was recovered.   In 1978, Ward and Matsumura extracted 2,3,7,8-TCDD from lake sediments with concentrations of 0.71, 1.0 or 1.83 ppm incubated from 1 hour to 589 days.   A sequential scheme of acetone, chloroform:methanol (4:1), hexane:acetone (5:1), and chloroform was used with recoveries in the range of 90 to 98 percent (Ward, 1978).   Both the Kearney, and Ward and Matsumura experiments showed a decrease in extraction efficiency with increased incubation periods, illustrating the increased difficulty of desorbing 2,3,7,8-TCDD with time (Philipp, 1981).   Finally, in 1980, Tiernan et al. found it necessary to use the Soxhlet extraction procedure to remove 2,3,7,8-TCDD from finely ground soil.   This is a lengthly procedure using pure methylene chloride for up to 3 days with recoveries of around 100 percent (Vanness, 1980).

Though 2,3,7,8-TCDD has a very low solubility in water (7 to 20 ppt) (Marple, 1986;   Adams and Blaine, 1986), micellar solubilization in a water medium has been attempted. Aqueous solutions of cationic, anionic, and nonionic solvents were tested in 1978 by Botre et al. on soil from Seveso, Italy.   The cationic solvent, 1-hexadecylpyridinium chloride (CPC) proved to be the best solubilizing agent of those tested, with 75 percent of the 2,3,7,8-TCDD on the soil being solubilized.   The other solubilizing agents tested were sodium dodecyl sulfate (SDS), methanol, and polyoxyethylene sorbitan monooleate (Tween 80).   Table 5.3.1 summarizes the results of these experiments.

5.3.3   Cost of Treatment

At the present time, due to the uncertainties involved in the use of this pretreatment technology, no definitive cost data have been made available. However, it is reportedly an expensive means of pretreating dioxin-contaminated soils because of the high costs associated with the solvents themselves and the high expected consumption rates needed for adequate extraction.

TABLE 5.3.1. SOLUBILIZATION OF TCDD (Botre, C., et al. 1978)

Solubilizer	Experiments on soil[a]		Experiments on pure TCDD[b]			
	Surfactant concn	Solubilized TCDD (%)	Surfactant conc	Solubilized TCDD (%)	Surfactant conc	Solubilized TCDD (%)
MeOH	-	97.5	-	100	-	100
CPC	0.05M	75.0	0.02M	75	0.05M	78
SDS	0.05M	60.0	0.02M	71	0.05M	75
Tween 80	2% w/v[c]	45.0	1% w/v[c]	72	2% w/v[c]	73

[a]Each 90-g sample contained initially 6.3 μg of TCDD.

[b]Initial amount of TCDD: 40.90 μg.

[c]Percent weight to volume

5.3.4 Process Status

More laboratory tests need to be made to determine the applicability of extraction procedures to the wide range of dioxin-contaminated soil types. For instance, high organic content soils are known to be harder to wash than sandy soils.

Though organic solvents have been identified with high removal efficiencies (in certain cases), the level of soil cleanup necessary will determine their applicability. Also, cleaning soils to 1 ppb or below has not yet been established as feasible. In addition, the solvent which is used in actual cleanup situations must be non-hazardous. Residual concentrations of solvents such as methylene chloride or carbon tetrachloride in soil would be unacceptable. Selection of the proper solvent also depends upon the final treatment or disposal scheme. The chemical CPC, for instance, is very sensitive to photodegradation (Botre, 1978).

Finally, pilot scale and full-scale testing is needed to resolve problems such as soil handling and preprocessing, as well as safety hazards involved in the increased mobility of 2,3,7,8-TCDD in the dissolved state.

5.4 BIOLOGICAL TREATMENT

5.4.1 Process Description

Biodegradation is the molecular breakdown of an organic substance by living organisms. During biodegradation, the decomposition which occurs results in less complex compounds which could be of either less or more toxicity. The principal factors which control microbial degradation are: moisture levels, organic content, oxygen levels, temperature, pH, and nutrient sources.

Biological treatment of wastes can be accomplished in a number of different modes. These include in situ aerobic degradation, pretreatment (e.g., by photolysis or ozonation) followed by biodegradation, anaerobic degradation, and activated sludge. Processes such as activated sludge, for which the waste must generally be in an aqueous form, are standard methods for treating domestic wastewater. For treating dioxin waste, which is frequently found in a soil matrix, in situ degradation is a more practical alternative. In addition to the various modes of treatment, biodegradation can be effected by a number of different types of micro-organisms. These include:

- aerobic bacteria;

- anaerobic bacteria;

- yeast; and

- fungi.

Following a discussion of the environmental degradation of TCDD, examples of research on the application of several of these modes of treatment and types of micro-organisms will be presented.

Degradation of 2,3,7,8-TCDD is a slow process, overall, in the natural environment. Natural degradation is primarily due to biodegradation and photochemical (UV) breakdown. A wide variety of half-lives have been reported. The observed half-life for uncontrolled biodegradation of 2,3,7,8-TCDD has been reported as 225 and 275 days by the U.S. Air Force (Young, 1976), although a separate analysis of the same data yielded half-lives ranging from 190 to 330 days (Commoner, 1976). Another study reported that half-life is affected by concentration, being greatly reduced at high concentrations (Bolton, 1978). In fact, half-lives are probably significantly greater than those reported, as most early research did not account for the strong tendency for 2,3,7,8-TCDD to bind to soil particles. Strongly bound 2,3,7,8-TCDD would not have been detected analytically and biodegradation assumed incorrectly to be the cause of its absence.

Studies at Seveso, Italy indicate that the half-life of 2,3,7,8-TCDD increases with its time in the soil, because of its tendency to become more tightly bound to soil and organic matter (DiDominico, 1980). DiDominico found that half-life calculations made 1-month after the Seveso accident predicted a 10 to 14-month half-life, but 17 months after the accident, the half-life of 2,3,7,8-TCDD in the soil had increased to more than 10 years.

In a study performed for the Air Force, 99 percent of the 2,3,7,8-TCDD sprayed as a constituent of defoliants was still present 12 to 14 years after application (Young, 1983). Although natural degradation seems to proceed fairly rapidly at first, it slows or completely stops after only a few months. This decrease in degradation is probably due to the affinity of 2,3,7,8-TCDD for soil and organic matter. Any use of biodegradation as a treatment process for 2,3,7,8-TCDD in soil will have to overcome this adsorption phenomenon.

Microbial metabolism of 2,3,7,8-TCDD has been shown to occur both in soil and aquatic environments. The species responsible for degradation were not always reported. Table 5.4.1 lists some of the microbial strains which are known to degrade 2,3,7,8-TCDD. Other researchers have identified soil micro-organisms with 2,3,7,8-TCDD degrading capabilities (Phillippi, 1982; Camoni, 1982).

The information available in the literature is incomplete with regard to the specific micro-organisms which have the capability for 2,3,7,8-TCDD degradation, their intermediate metabolites, and end products of biodegradation.

The process of 2,3,7,8-TCDD biodegradation currently appears to be one of co-metabolism. That is, 2,3,7,8-TCDD is not metabolized directly as a food or energy source, but is degraded by metabolic enzymes generated during the metabolism of other organics. The degree of 2,3,7,8-TCDD biodegradation which can occur has been reported to vary for different micro-organisms.

5.4.2   Technology Performance Evaluation

Past Research--

Research on the degradation of 2,3,7,8-TCDD has been going on for a number of years. Work prior to 1980 monitored various aspects of biodegradation in the soil environment. Much of this early work is subject to dispute because biodegradation was inferred from the "disappearance" of 2,3,7,8-TCDD. It is now known that 2,3,7,8-TCDD binds tightly to soils and that biodegradation rates are much lower than early reports indicated. Other researchers followed the metabolism of carbon 14 labeled 2,3,7,8-TCDD or measured the formation of 2,3,7,8-TCDD metabolites. The following discussion summarizes a literature review of past research on soil biodegradation of 2,3,7,8-TCDD (Esposito, 1980).

TABLE 5.4.1.   MICRO-ORGANISMS WITH KNOWN CAPABILITY FOR
DEGRADING 2,3,7,8-TETRACHLORODIBENZO-P-DIOXIN

	Researcher
Nocardiopsis sp.	Matsumura, 1983.
Bacillue megaterium	Matsumura, 1983.
Beijerinckia B8/36[a]	Klecka, 1980.
Pseudomonas, sp.[b]	Klecka, 1979.
Biejerinckia, sp.[a]	Klecka, 1980.
Phanerochaete chrysosporium[c]	Bumpus, et al., 1985.

[a]Oxidation of dibenzo-p-dioxin and several mono-, di-, and
trichlorinated dibenzo-p-dioxins was reported.

[b]Metabolism of dibenzo-p-dioxin was observed.

[c]White rot fungus.

A 1973 study by researchers at the Agricultural Research Center in Beltsville, Maryland monitored 2,3,7,8-TCDD degradation in two soils with 2,3,7,8-TCDD concentrations of 1, 10, and 100 ppm, and with ^{14}C-labeled concentrations of 1.78, 3.56, and 17.8 ppm (Kearney, 1972). Soil samples were monitored for evolution of $^{14}CO_2$ as an indication of microbial degradation, but very little was detected and no metabolites were found in 2,3,7,8-TCDD treated soil after 1-year.

Camoni, et al., investigated the influence of organic compost additions to 2,3,7,8-TCDD-contaminated soils from Seveso, Italy. The organic compost was used to provide active micro-organisms to stimulate biodegradation. The researchers concluded that organic compost addition had little effect on the degradation of 2,3,7,8-TCDD. Initial concentrations of TCDD in soil were 100 ppb. At the end of the experiment, extracted 2,3,7,8-TCDD was about 73 percent of that extracted initially for the compost treated soil, and 88 percent of initial levels for untreated soils.

Hutter and Phillippi of the Microbiological Institute of Zurich, Switzerland found aerobic microbial degradation of 2,3,7,8-TCDD in liquid media or soil to be very low under laboratory conditions. Approximately 1 percent of the input material was degraded after several months of incubation to an apparent hydroxlated 2,3,7,8-TCDD metabolite. Initial 2,3,7,8-TCDD concentrations were not reported.

Ward and Matsumura (1978) from the University of Wisconsin studied the fate of TCDD using aquatic sediment and lake water. Under experimental conditions the half-life of TCDD was approximately 600 days. Maximum levels of metabolite production were reached between 19 to 39 days of incubation. The level of metabolite production amounted to 1 to 4 percent of the original TCDD level.

Other researchers who have investigated biodegradation of 2,3,7,8-TCDD include: Bartleson, et al., 1975; Commoner and Scott, 1976; Pocchiari, 1978; and Salkinoja-Salonen, 1979. As noted, Esposito, et al., 1980, has summarized the work published by these researchers. The indications are that biodegradation will proceed slowly in soils and may be stimulated by addition of nutrients and organics.

Ongoing Research--

There have been several more recent research projects concerning the biodegradation of 2,3,7,8-TCDD and related compounds. Some of the more significant ones are discussed below.

White Rot Fungus (Bumpus et al., 1985)--

One method that has received a large amount of attention has involved the study of the ability of the fungus, P. chrysosporium, to degrade recalcitrant organopollutants, one of these being 2,3,7,8-TCDD. P. chrysosporium is a lignin-degrading white rot fungus. This organism secretes a unique hydrogen peroxide-dependent oxidase capable of degrading lignin, a highly complex, chemically resistant, nonrepeating heteropolymer. The enzyme catalyzes the formation of carbon-centered radicals which react with oxygen to initiate oxidation. The low molecular weight aromatic compounds formed may then undergo further modification or ring cleavage and eventually be metabolized to carbon dioxide.

Several properties of P. chrysosporium make it a candidate for the degradation of the more recalcitrant organopollutants such as 2,3,7,8-TCDD, DDT, lindane and PCBs. First of all, the organism is able to degrade lignin, chlorinated lignin and chlorinated lignin-derived by-products of the Kraft pulping process. Secondly, low levels of pollutant (such as may exist in contaminated soil) do not repress the production of enzymes required for degradation. Thirdly, the organism is not substrate-specific and therefore can attack and degrade a wide variety of structually diverse, recalcitrant compounds. Finally, P. chrysosporium is a highly successful competitor in nature, especially when the carbon source is lignin. Consequently, competition by other organisms will be minimal if wood chips or sawdust are added as a supplement to the waste material.

Results of laboratory tests using P. chrysosporium to degrade several different compounds are shown in Table 5.4.2. In 10 ml cultures containing 1.25 nmoles of the ^{14}C-labeled 2,3,7,8-TCDD substrate, 27.9 pmoles were converted to ^{14}CO$_2$ within 30 days and 49.5 pmoles within 60 days, representing 4.96 percent metabolism. The remaining carbon atoms should have either been incorporated into the organism or been present as intermediates in the pathway between 2,3,7,8-TCDD and CO$_2$. This conclusion is based on more

TABLE 5.4.2.   DEGRADATION OF ^{14}C-RADIOLABELED ORGANOPOLLUTANTS TO $^{14}CO_2$
BY P. CHRYSOSPORIUM (Bumpus, et al., 1985)

	Initial rate of degradation to $^{14}CO_2$ (pmoles/day)	Radiolabeled substrate evolved as $^{14}CO_2$ (pmoles)		% of Radiolabeled substrates evolved as $^{14}CO_2$ in 60 days
		30 days	60 days	
Lindane	11.3	190.8	267.6	21.4
Benzo(a)Pyrene	7.5	117.2	171.9	13.8
DDT	2.7	48.0	116.4	9.3
TCDD	1.2	27.9	49.5	4.0
3,4,3',4'-TCB	0.7	13.8	25.1	2.0
2,4,5,2',4',5'-HCB[a]	2.4	44.2	86.0	1.7

[a]Substrate concentration was 1.25 nmoles/10 mL for all ^{14}C-radiolabeled compounds except 2,4,5,2',4',5'-HCB. Because of its low specific radioactivity a concentration of 5.0 nmoles/10 mL was used for 2,4,5,2',4',5'-HCB.

detailed studies of the degradation of DDT which indicated that after 30 days 4 percent of the original DDT was evolved as CO_2, when approximately 50 percent of the DDT had been degraded. In the case of DDT, greater than 99 percent degradation had occurred after 75 days.

Matsumura and Quensen (Quensen and Matsumura; 1983, Quensen, 1986)--

Research has been conducted at the Pesticide Research Center of Michigan State University in which low concentrations of 2,3,7,8-TCDD were metabolized by pure cultures of Nocardiopsis spp. and Bacillus megaterium. In these experiments TCDD in solvent was added to flasks containing the pure cultures, and after a period of 1 week, the contents were extracted and analyzed for TCDD and metabolites.

Several conclusions were drawn from the study. One of these, as substantiated by detail in Table 5.4.3, is that the choice of solvent used to dissolve TCDD and add it to the culture medium has a significant effect on the degradation of TCDD. The use of ethyl acetate or dimethyl sulfoxide (DMSO) resulted in significantly higher degradation than when corn oil or ethanol were used. Another conclusion that was drawn is that lowering of alternative carbon sources increases the degradation of TCDD. The proportion of TCDD metabolized by B. Megaterium increased dramatically when the amount of soybean extract in the medium was reduced from 1.6 to 0.4 percent and ethyl acetate was used as the solvent. Finally, analog-induced metabolism of TCDD by including napthalene or dibenzofuran in the culture medium proved to be ineffective.

In addition to the pure culture experiments, TCDD degradation in soil by naturally occuring micro-organisms was also studied. TCDD was added to three different types of soil and after 0, 2, 4, and 8 months of incubation, soil samples were extracted and the levels of TCDD and metabolites were determined. Very little metabolism of TCDD occured in any of the soils over the 8 month period. This was true regardless of which solvent was used to add TCDD to the soil system. Dimethyl sulfoxide, ethyl acetate and 10 percent ethanol were used. They surmised that the resistance to degradation was due to the fact that TCDD binds tightly to soil thereby limiting the rate of cellular uptake.

TABLE 5.4.3. EFFECT OF SOLVENTS ON METABOLISM OF TCDD BY BACILLUS MEGATERIUM
IN YEAST-SOYBEAN MEDIUM (Quensen and Matsumura, 1983)

| | Amount of radiocarbon recovered (%) | | | | |
| | Aqueous phase | | Solvent phase | | |
Treatment	A[a]	B[b]	Metabolites	TCDD	Total
Ethyl acetate	4.3 ± 1.3	12.4 ± 16.5	8.9 ± 3.5	51.8 ± 27.5	77.4 ± 15.6
DMSO	1.9 ± 1.4	1.1 ± 0.5	8.9 ± 0.9	81.9 ± 13.1	93.7 ± 10.3
Ethanol	5.9 ± 2.3	1.4 ± 1.6	1.6 ± 0.2	90.0 ± 6.6	98.8 ± 3.0
Corn oil	0.2 ± 0.2	1.0 ± 1.0	c	92.0 ± 1.1	93.2 ± 2.3

[a]Extracted medium.

[b]Aqueous layer formed during evaporation of solvents.

[c]Thin-layer chromatographic analysis was not possible because of interference
by the corn oil.
The amount of solvent used was 1 mL per 50 mL culture. Values are means ± SD;
sample size = 2.

<u>Kearney and Plimmer</u> (Kearney, 1984)--

A biological process to detoxify 2,3,7,8-TCDD-contaminated soils is being evaluated at the Agricultural Research Center in Beltsville, Maryland. Work is based on the observation that soil micro-organisms have the ability to degrade highly chlorinated organics that have been pretreated with ultra-violet (UV) radiation. Pretreatment with UV radiation removes chlorine from the 2,3,7,8-TCDD molecule in the presence of a proton donor, and the resulting dibenzo-p-dioxin molecule can then be biodegraded.

Laboratory studies have involved subjecting solutions of chlorinated organics to UV radiation before adding them to the soil where biodegradation could take place. Kearney and his colleagues have since experimented with a prototype system that includes a 55-gallon stainless steel drum as a holding tank and a commercial water purifying unit as the UV source. Kearney's process focuses on cleaving the chlorine-carbon bonds in the chlorinated organic compounds by the following procedures:

1.   Expose a dilute aqueous solution (i. e., 1 ppm 2,3,7,8-TCDD) to ultra-violet light (UV) for at least 1 hour (Photo-Chemical Reaction).

2.   While irradiating, bubble oxygen through the solution to speed up the chlorine-carbon bond break up (Ozone Reaction).

3.   Pour or spray irradiated solution over soil containing the test micro-organisms (Biodegradation).

4.   Determine the percent degradation by monitoring the amount of carbon dioxide generated.

Recent studies (1981) have yielded the following results:

- 80 percent degradation of 2,4,5-T over 1 month;

- 80 percent degradation of PCB over 1 month;

- 60 percent degradation of 2,3,7,8-TCDD over 1 month; and

- 75 percent degradation of PCP over 1 month.

For all compounds, most of the decomposition occurred during the first day or two in the soil.

This process is limited in applications to dioxin molecules that are in solution with a proton donor and is most effective when 2,3,7,8-TCDD concentrations do not exceed 1 ppm. The UV pretreatment step cannot be applied to 2,3,7,8-TCDD-contaminated soil because UV radiation does not penetrate below the soil surface.

Loper (Poiger, 1983; Loper, 1985)--

Research is being conducted on yeast at the University of Cincinnati. Loper hopes to genetically alter yeast to include a gene for a liver enzyme (p450 mono-oxygenase) that is able to degrade dioxin molecules. Degradation of dioxin molecules has been observed in dogs and rats due to liver enzymes. Results are not available at this time.

Research on TCDD Surrogates--

Compounds such as chlorophenols, chlorobenzenes and the herbicides 2,4,5-T and 2,4-D are structurally similar to chlorinated dioxins, and therefore micro-organisms that have the ability to degrade these compounds may have the ability to degrade TCDD as well. In addition, since TCDD is a byproduct of manufacturing chlorophenols and 2,4,5-T, these compounds are frequently found at waste sites along with TCDD. Several groups have conducted research on the biodegradation of these types of compounds. These research projects are summarized in Table 5.4.4. Most of these projects are currently only at the laboratory scale and many of them are not necessarily directed toward the treatment of waste containing TCDD. Nonetheless, further research may indicate that TCDD can be degraded by these processes.

TABLE 5.4.4.   SUMMARY OF RESEARCH PROJECTS ON BIODEGRADATION OF TCDD SURROGATES

Researcher	Compounds tested	Type/Name of micro-organism	Type of process	Performance	Reference
University of Illinois Medical Center	2,4,5-T;2,4-D, Chlorophenols; research has been funded for TCDD	Pseudomonas Cepacia, AC1100	micro-organisms developed by acclimation to 2,4,5-T in a chemostat; bacteria also can be applied to soil	up to 98% degradation of 2,4,5-T; reduction of 2,4,5-T in soil from 1000 ppm to 30 ppm in one week	Ghosal, et al., 1985; Kilbane et al., 1983; Tomasek and Chakrabarty, 1985
Michigan State University	2,4,5-T, chlorinated phenols	unidentified anaerobic bacteria	would probably involve the use of an anaerobic digester	capable of removing chlorine atoms from the chlorobenzoate molecule	Tiedje, 1984
Sybron Corporation, Salem, Virginia	3,4-Dichlorophenol	Pseudomonas Stulzeri	addition of microbes in powdered form to contaminated site	lab testing has shown degradation at 3,4-DCP concentrations of 50-100 ppm	Davis, 1984; Goldsmith, 1986
Groundwater Decontamination Systems, Inc., Waldwick, N.J.	acetone; methylene chloride; n-butyl alcohol; 1,4-dichloro-dibenzo-p-dioxin (1,4-DCDD)	unknown	activated sludge system from which micro-organisms are injected into groundwater	600-700 ppm of contaminant reduced to less than 6 ppb for industrial organics	Macazza, 1983
Louisiana State University, Hazardous Waste Research Center	2,4-D and 2,4,5-T	Pseudomonas (only identifiable at the genus level), Alcaligenes eutrophus	laboratory scale batch reactors	demonstrated growth of micro-organisms using 2,4-D as sole carbon source	Roy and Mitra, 1986
University of Minnesota	2,4-D; 3,5-dichlorobenzene (3,5-DCB)	sewage treatment plant sludge was the source of innoculating after acclimation, the Pseudomonas species was predominant	laboratory batch reactors and chemostats	micro-organisms were capable of utilizing 2,4-D and 3,5-DCB as the sole substrate at concentrations between 10ug/l and 100mg/l	Kim and Maier, 1986

5.4.3 Costs of Treatment

Costs associated with biological treatment of TCDD will undoubtedly be expensive, but given the high cost associated with other treatment alternatives, biological treatment is expected to be very competitive. In-situ treatment would be the least expensive method. Operations would be similar to an intensive agricultural process, with regular tilling, fertilizing, and irrigation. Adaption of activated sludge to treatment of 2,3,7,8-TCDD-contaminated soils would be much more expensive. If the soil were excavated and suspended in an aeration tank, aeration costs would be very high. Activated sludge would be most cost-effective if applied to 2,3,7,8-TCDD which had been solvent extracted from the soil.

5.4.4 Process Status

At this point in time, biodegradation as a process for treating wastes containing TCDD is still very much in its infancy. Certain types of micro-organisms have shown the ability to degrade TCDD in laboratory cultures. With respect to actual wastes such as contaminated soil, biological treatment has not been demonstrated to be effective. The major obstacle has been that it has been difficult to make TCDD bioavailable. Since TCDD adsorbs strongly to soils, cellular uptake is severely limited. One organism discussed above, the white rot fungus (P. Chrysosporium), has the ability to degrade recalcitrant insoluble substrates such as lignin by secretion of extracellular enzymes. Because of this ability, it has been suspected that this fungus might have the ability to degrade 2,3,7,8-TCDD and other haloorganics which are resistant to degradation by other micro-organisms. Some initial data have indicated that P. Chrysosporium has a potential for degrading TCDD to nontoxic end products. Even these data, however, are inconclusive, since only 5 percent of the degradation end products were measured as CO_2 after 60 days. Much additional work must be carried out before P. Chrysosporium can be used to treat actual waste. Research will be carried out in the near future to determine the survivability of the micro-organism in a soil environment. In a lab-scale study, moisture, temperature, pH, and wood chip type and size (the substrate on which the micro-organism is grown) will be varied to determine optimum growth conditions (Sferra, 1986). After this, the white rot fungus will be tested in soil plots containing TCDD. This means that the application of even a promising micro-organism such as P. Chrysosporium to actual waste will not occur for several years, and only then if it proves to be successful in lab-scale testing.

Several treatment methods described above have been demonstrated to be applicable to the treatment of compounds that are similar to 2,3,7,8-TCDD, such as chlorophenols and 2,4,5-T. They have not, however, been tested either in the lab or the field on waste containing TCDD. In the case of the Sybron Corporation work, it has been difficult to obtain samples of waste with which to test their process in the lab, and it has also been difficult to test their process at actual sites of TCDD contamination (Goldsmith, 1986). One of the reasons for the difficulty in testing microbial processes on actual waste sites is the issue of releasing genetically altered micro-organisms to the environment.

In summary, the feasibility of biodegradation of 2,3,7,8-TCDD as a treatment technology is still in question. Most investigations have been performed in the laboratory, and the efficiency of a large scale treatment process is unknown. There are many advantages associated with biotreatment which make continued investigation advisable:

- The end products of complete biodegradation are nontoxic.

- Some processes may be accomplished onsite without soil excavation. However, use of solvents which could potentially cause uncontrolled mobilization of 2,3,7,8-TCDD must be avoided.

- Biological treatment appears to be effective at low 2,3,7,8-TCD concentrations.

In addition, biotreatment could be coupled with other treatment processes to make them both more efficient. For instance, the sodium polyethylene glycol (NaPEG) process might be modified to in-situ treatment with the use of micro-organisms to degrade the dechlorinating solvent and the residual nonchlorinated and less chlorinated dibenzo-p-dioxins.

5.5   STABILIZATION/FIXATION

5.5.1   Process Description

Stabilization/fixation processes are used to render immobile the
hazardous constituents which may be present in liquids, sludges, or solids.
Stabilization refers to processes which physically change or encapsulate the
waste through mixing with additives and binders. Fixation involves a chemical
interaction between the waste and a chemical binding agent. Stabilization
techniques are only interim measures. The stabilized wastes must be
subsequently disposed at an EPA-approved hazardous waste landfill, or
subjected to further treatment.

Several types of stabilization/fixation technologies exist, including:
cement-based, lime-based (pozzolanic), thermoplastic and thermosetting organic
polymers, macroencapsulation, self-cementing, glassification, and other more
recently developed techniques (Tittlebaum, et al., 1985). Table 5.5.1 lists
some of the stabilization/fixation techniques in these categories. Most must
be considered stabilization techniques; fixation processes are rare.

Stabilization materials can be either organic or inorganic; inorganics
are more commonly used and include Portland cement, pozzolanic materials with
or without lime or cement, and sorbent clays (Hazardous Waste Consultant,
1985; Spooner, 1985; GCA, 1985). Organic materials include asphalt,
polyethylene, urea formaldehyde, and other thermoplastic and thermosetting
polymers (Hazardous Waste Consultant, 1985; Spooner, 1985; GCA, 1985). The
performance of most of these materials over time has not been fully
demonstrated.

Organic Materials--

One common technique for stabilizing organic contaminants is blending
them into resins and then solidifying the mixtures (GCA, 1985). Plastic
solidifying agents fall into two main categories, thermoplastics and
thermosets. Thermoplastics are materials that become fluid upon heating and
include asphalt, polyethylene, polypropylene, and nylon. Thermoplastic
techniques generally call for the waste to be dried, heated, dispersed through
the heated plastic matrix, and then cooled (solidified) and placed in

TABLE 5.5.1   SUMMARY OF STABILIZATION PROCESSES FOR TREATING
HAZARDOUS WASTES (M. E. Tittlebaum, et al., 1985)

Classification	Process sponsor (process name)	Stabilization agents	Wastes treated
Cement-based	Chemfix	Cement, soluble silicates	Inorganics
	Stablex (Sealosafe)	Cement, flyash	Inorganics
	Stabatrol (Terra-Tite)	Cement, additives	Inorganics
Lime-based or pozzolanic	Dravo Lime (Calcilox)	Lime, additives	FGD sludges
	International Mill Service	Lime	Metal slags
	IU Conversion Systems	Lime	FGD sludges
	Soil Recovery Systems	Lime	Misc.
	Sludgemaster	Lime, additives	Misc.
Thermoplastic	Werner and Pfleiderer	Asphalt	Misc.
	Southwest Research Inst. (Sulfex)	Sulfur, modifiers	Misc.
Thermosetting polymer	Dow Chemical	Polyesters, polyvinyls	Radioactive
	Newport News Industrial	Polyesters	Radioactive
Macro-encapsulation	Environmental Protection Polymers	Polyolefins	Soluble toxics
	TRW Systems	High-density polyethylene	Misc.
Self-cementing	Sludge Fixation Technology (Terra-Crete)	Calcium sulfite or sulfate	FGD sludges
Glassification	None specified	Glass or ceramics	Radioactive
Other	ARDECCA	Proprietary	Oil field wastes
	Anschutz Corp. (Ansorb)	Clay-like material	Misc.

containers. Thermosets include urea formaldehyde, polyester, and phenolic and melamine resins. Thermoset techniques call for the waste to be mixed with the thermoset prior to reaction of the mixture to form a solid matrix through crosslinking reactions. This matrix will remain solid throughout subsequent heatings. Containers may or may not be needed with thermosets.

In early work, asphalt and bitumen were the most widely applied materials for solidifying organics (GCA, 1985). These fixative materials are chemically stable and low in cost. At low waste-to-fixative loadings, these materials were generally found to exhibit acceptable solidification properties (e.g., good solid product formation and dimensional stability remained upon immersion in water). However, for high contaminant loadings (above about 30 percent by weight), or, in general, for most organics of lower molecular weight, high vapor pressure or hygroscopic nature, these materials often yielded unacceptable products. More recently, these products have been replaced by thermoplastic or thermosetting resins; e.g., linear polyethylene has been employed as a stabilizer for certain organics (GCA, 1985).

Soil stabilization chemicals are also available that react with moisture in the soil or an aqueous catalyst to form a hydrophobic crosslinked polymer-based gel (GCA, 1985). The semi-solid gel forms in situ coats and binds the soil particles together. The chemical and water (or catalyst) mixture is sprayed on cultivated or loosened soil to react with the upper 3 to 4 inches of soil. The resulting gel-soil mixture then becomes a barrier to water infiltration.

Commonly offered grouts include organic polymers based on acrylamides, polyurethanes, urea, and phenolics. The advantages some of the chemical grouts offer are that they are easy to mix, they penetrate soil much like water (since they are polar and have a viscosity similar to water), they can be applied by spraying, and they are generally nontoxic when handled properly. The grouts form highly stable compounds of extended but unknown life. However, grouts are sensitive to freeze-thaw and wet-dry conditions, and some grouts will deteriorate under ultraviolet light and other degradative mechanisms.

Inorganic Stabilizers--

Lopat Enterprises, Inc., of Asbury Park, New Jersey has developed a product called "K-20" (McDaniel, 1983). K-20 is an inorganic mixture of at least eight chemicals (McDaniel, 1983). No further information on its content is available at this time. Lopat, Inc., has a patent pending for K-20, which was originally developed as a sealant for leaky basements (Goldensohn, 1983). Recent investigations have shown evidence that dechlorination of chlorocarbon contaminants may occur when K-20 is used as an encapsulant (Jiranders, 1984).

Pretreatment Requirements/Restrictive Waste Characteristics--

Stabilization is more frequently used for inorganics because organics tend to interfere with the physical and chemical processes which are necessary to bind the materials together (Spooner, 1985; Hazardous Waste Consultant, 1985; GCA, 1985). Wastes with greater than 10 to 20 percent organic content are generally not recommended for treatment by stabilization.

5.5.2 Treatment Performance Evaluation

In 1985, the Solid Waste Research Division of the Disposal Branch of the U.S. EPA sponsored a study to find the optimum mixture of asphalt and soil cement that will stabilize 2,3,7,8-TCDD-contaminated soil (Vick, 1985). The Portland Cement Institute and the Asphalt Institute will be reviewing the work. In the laboratory, stabilized soil underwent a leach test designed by Battelle- Columbus Laboratory. A structural integrity test was suggested, but not undertaken, because the soils are not expected to be subject to large loads (even though strength tests are often an indication of durability).

JRB Associates, under the sponsorship of the U.S. EPA, conducted a field test of cementious and asphaltic stabilization techniques in the State of Missouri during 1985 (Vick, 1985; Ellis, 1986). The objectives of this testing program included:

● evaluating the cost-effectiveness of the processes;

● developing optimum soil/stabilizer ratios and mixing conditions; and

● assessing the viability of successful field implementation.

Soil from three Missouri dioxin sites were tested with both stabilization techniques. As summarized in Table 5.5.2, each soil had different properties. Cement-based stabilization samples were prepared at optimum moisture, but with varying cement contents. Tests were performed for freeze-thaw susceptability, leachability, and 7-day unconfined, comprehensive strength (Vick, 1985). The freeze-thaw and wet-dry test results were satisfactory for the portland cement process. However, significant loss (by weight-percentage) of the sample was observed during weathering processes followed by an aqueous leaching process. Percent weight loss in the soil/cement samples ranged from 6 to 18 percent for the Minker sample, 13 to 16 percent for the Piazza Road sample, and 16 to 27 percent for the Sontag Road sample (Vick, 1985; Technical Resources, 1985). The leachability test results are summarized in Table 5.5.3. The results suggest that the leachate concentration may be limited by the decreased aqueous solubility of 2,3,7,8-TCDD in the range of 2 to 3 ppt in the leachate from the matrix (Technical Resources, Inc., 1985).

A cationic, slow-setting emulsion (CSS-1h) was used for the emulsified asphalt samples. A slow setting emulsion is preferred for dense-graded aggregate soils such as the Missouri soils used during these tests. Initial tests for wet-dry stability, dry density, moisture content, and air void, indicated that the emulsified asphalt did not produce acceptable results (Vick, 1985; Ellis, 1986). Therefore, formula modifications using lime additives (calcitic lime, calcium hydroxide) were developed to modify the soil prior to asphalt addition (Ellis, 1986). Addition of 1.5 percent lime and the use of the SS-1h (nonionic, slow-setting emulsion) instead of the CSS-1h emulsified asphalt, dramatically improved the results of the soil/asphalt sample tests (Vick, 1985). The optimum asphalt percentage was found to be nine percent residual asphalt (Vick, 1985).

Lopat, Inc. has recently completed tests of the K-20 process on contaminated soils from Times Beach. The data from these tests have not been of sufficient quality to permit an accurate assessment of the process with respect to the encapsulation of dioxin. Lopat plans to conduct further tests on dioxin-contaminated soils when samples are made available (Flax, 1986).

TABLE 5.5.2.   SOIL TYPES USED TO TEST PORTLAND CEMENT
AND EMULSIFIED ASPHALT/LIME STABILIZATION
TECHNIQUES

Site	Description	Soil type	TCDD concentration
Minker	Residential area with steep, sloping banks that drain into a nearby creek	Sandy loam	700 ppb
Piazza Road	Roadside material	Sandy loam	640 ppb
Sontag Road	Roadside material, with considerably greater percentage of fine particles (salt and clay) than Piazza Road sample	Sandy silty loam	32 ppb

(Technical Resources, Inc., 1985).

TABLE 5.5.3.   SUMMARY OF LEACHABILITY TEST RESULTS FOR PORTLAND CEMENT
STABILIZED AND NATIVE UNSTABILIZED MISSOURI SOILS[a]
(Technical Resources, Inc., 1985)

Soil collection site	Material leached	Mass TCDD leached (ng)	Conc. TCDD in leachate (ng/L)	Mass TCDD leached per unit mass of soil (ng/g)
Minker	Unstabilized	9.5 $\pm$ 4.7	2.4 $\pm$ 1.2	0.095 $\pm$ 0.047
Minker	Soil/cement	28 $\pm$ 8	2.3 $\pm$ 0.7	0.012 $\pm$ 0.003
Piazza	Unstabilized	3.9	1.0	0.039
Piazza	Soil/cement	36 $\pm$ 12	2.6 $\pm$ 0.9	0.016 $\pm$ 0.006
Sontag	Unstabilized	3.4	0.91	0.034
Sontag	Soil/cement	8 $\pm$ 4[b]	0.6 $\pm$ 0.4	0.004 $\pm$ 0.002

[a]Mean values of triplicate measurements, unless indicated.

[b]One of the three replicates was below detection limit; listed value is a
mean of two measurements.

The University of Maryland is currently performing controlled tests on the ability of K-20 to decontaminate soils by encapsulation and/or dechlorination using several chlorinated hydrocarbons including 2,3,7,8-TCDD. The U.S. EPA in Cincinnati, Ohio is also running some tests on the ability of K-20 to degrade 2,3,7,8-TCDD in soil. The results of the last two sets of tests may help to evaluate the effectiveness of this encapsulation agent.

5.5.3   Costs of Treatment

To date, stabilization/fixation processes have not been fully tested and cost effectiveness has not been documented. Organic wastes are generally more practically disposed of via other technologies such as incineration. Fixation becomes more cost-effective when the organic content of the waste is small, thereby making incineration less feasible (e.g., 2,3,7,8-TCDD-contaminated soils).

5.5.4   Process Status

Application of stabilization/fixation processes to organic wastes is a relatively recent development, because organic wastes generally lend themselves better to other treatment processes such as incineration or biodegradation. Preliminary studies of contaminated soil suggest that an emulsified asphalt-lime combination may be an effective interim remedial measure for stabilization of dioxin-contaminated soils (Vick, 1985; Ellis, 1986).

Further studies plan to investigate the leaching potential and performance of formulations using uncompacted soils (Vick, 1985). Future goals are to develop a procedure whereby temporary in-situ stabilization could be followed by soil/stabilizer removal and complete stabilization or fixation at an offsite facility.

5.6   CHEMICAL DEGRADATION USING RUTHENIUM TETROXIDE

5.6.1   Process Description

Ruthenium tetroxide is a powerful oxidizing agent. It is more effective than either hypochlorite or permanganate in attacking aromatic substances. The reagent can be used in solution with water or with organic solvents which demonstrate no nucleophilic character such as chloroform, methylene chloride, acetic acid, fluorotrichloromethane, and nitromethane.

Degradation using ruthenium tetroxide is by aromatic ring cleavage. In tests where chlorophenols were treated with ruthenium tetroxide, all of the aromatic ring carbons were accounted for as carbon dioxide, and the aromatic chlorosubstituents gave rise to chloride ions. A similar analysis of the degradation products of TCDD was not carried out in this study due to analytical difficulties related to the low solubility of the compound. It was inferred, however, that because of the close chemical and structural similarities between TCDD and chlorophenols that they would be degraded in a similar manner (Ayres, 1985).

One factor affecting the rate of pollutant degradation using ruthenium tetroxide (RuO_4) is temperature. In one set of experiments, the rate of pollutant degradation increased 2.4-fold per 10 °C rise within the test temperature range (Ayres, 1981a; des Rosiers, 1983).

5.6.2   Technology Performance Evaluation

Studies have been performed on soils contaminated with Agent Orange. These soil samples, containing approximately 70 ppb of TCDD, were obtained from Eglin Air Force Base in Florida. After treatment of the extracted material with excess ruthenium tetroxide for 1 hour at 76°C in carbon tetrachloride, TCDD was no-longer detectable (detection limit of 10 ppb). When water was used as the solvent instead of CCl_4 the degradation of TCDD was noticable, but not nearly as great as with CCl_4. An experiment was also performed in which an excess of sodium hypochlorite and hydrated ruthenium tetroxide were added to the same soil samples along with carbon tetrachloride as the solvent. In this case, the degradation was similar to that attained by using excess ruthenium tetroxide in the presence of CCl_4 (Ayres, 1985).

Tests have also been done on synthetic wastes containing PCDDs. In one of these experiments, 2,7-DCDD, when mixed with RuO_4 in a carbon tetrachloride solution was determined to have a half-life of 215 minutes at 30°C; the half-life decreased to 38 minutes at 50°C. The oxidation of 2,3,7,8-TCDD proved to be a slower reaction; at 20°C it had a half-life of 560 minutes, while at 70°C the half-life decreased to slightly less than 15 minutes.

5.6.3   Costs of Treatment

Due to the current level of development of this technology, no cost data are available. Major costs would be for energy to heat up the material to be treated, and the cost of the chemical reagents. Pretreatment, extraction and post-treatment costs are unknown.

5.6.4   Process Status

To date, this method of degrading TCDD has only been performed on a laboratory scale. While these studies have shown that RuO_4 has the ability to degrade 2,3,7,8-TCDD, the reaction end products have not been identified. In addition, the only work reported has involved either the use of water or CCl_4 as the solvent. Water is not very effective, and the application of carbon tetrachloride to soil would not be environmentally acceptable. Thus, the use of other solvents should be investigated.

This technology will require considerable work before it can be applied in the field. The high cost of ruthenium tetroxide and the toxicity of process residuals may limit application of this technology. Its potential (if any) probably lies in the area of detoxification of glassware or purging of industrial reactors (des Rosiers, 1986).

5.7   CHEMICAL DEGRADATION USING CHLOROIODIDES

5.7.1   Process Description

A method for the degradation of substances containing both aromatic rings and ether bonds was reported in 1979 (Botre, 1979; des Rosiers, 1983; Esposito, 1980). This is of current interest because 2,3,7,8-TCDD contains

two aromatic rings connected by two ether bonds. The method utilized
chloroiodides attached to quarternary ammonium salt surfactant molecules to
rupture the ether bonds, and thus split the 2,3,7,8-TCDD molecule into smaller
fragments. End products are chlorophenols and related compounds. The
mechanism of the ether bond rupturing is thought to be the loss of an iodine
atom from the surfactant (Corwell, 1957), and subsequent formation of reactive
hydrogen iodide at a location in an aqueous solution near the 2,3,7,8-TCDD
molecule (Botre, 1979). Hydrogen iodide by itself is known to rupture ether
bonds, but usually only in a strongly acidic environment. However, the
formation of the hydrogen iodide in close proximity to the 2,3,7,8-TCDD
molecule seems to be the key factor.

One method of 2,3,7,8-TCDD degradation described involves the extraction
of 2,3,7,8-TCDD from soil using aqueous solutions of surfactants containing
chloroiodide groups (Botre, 1979; des Rosiers, 1983, Esposito, 1980). The
aqueous residues from the soil washings are extracted with benzene, methanol,
or methylene chloride. These extracted liquids containing the 2,3,7,8-TCDD
may require evaporation under reduced pressure to concentrate the solution and
thus enhance the reaction by bringing the 2,3,7,8-TCDD molecule and
chloroiodide-bearing surfactant molecules into more frequent contact.

The chloroiodide derivatives producing the most promising results for the
cleavage of ether bonds are alkyldimethylbenzylammonium (benzalkonium)
chloroiodide, and 1-hexadecylpyridinium (cetylpyridinium) chloroiodide (CPC).
The low solubilities of these chloroiodides in water can be increased with the
addition of micellar solutions of the same surfactants with chloroiodide
groups. Micellar solutions consist of large polymeric particles (clusters) of
the surfactants. Common solubilizing agents are benzalkonium chloride, used
to enhance the solubility of benzalkonium chloroiodide, and cetylpyridinium
chloride, used to enhance the solubility of cetylpyridinium chloroiodide.

Surfactant micellar solutions of 2,3,7,8-TCDD without chloroiodides are
stable when stored in the dark, but decompose when exposed to sunlight or UV
irradiation (Botre, 1978). This form of treatment seems to be appropriate and
effective for the decontamination of buildings, furniture, etc., where
surfactant contact accompanied by exposure to UV is possible (Botre, 1978).
However, the use of chloroiodides has been shown to be effective in the

decomposition of 2,3,7,8-TCDD without irradiation. This latter method utilizing chloroiodides is therefore more suitable for degrading bulk solutions.

5.7.2   Technology Performance Evaluation

No commercial processes utilizing chloroiodides for decomposition of 2,3,7,8-TCDD are known to exist. However, experiments illustrating the use of surfactants containing chloriodides for the cleavage of ethers have been accomplished. These experiments have been performed on substances such as xanthene, benzofuran, and 2,3,7,8-TCDD (Botre, 1979). All substances tested confirmed that chloroiodides aided in the decomposition. This discussion will be limited to the results from experiments on 2,3,7,8-TCDD.

In one study, solutions containing 2,3,7,8-TCDD in benzene were vacuum evaporated and the residues were treated with aqueous surfactant solutions (Botre, 1979). Two chloroiodide derivatives were used in the surfactant solutions: benzalkonium chloroiodide, and cetylpyridinium chloroiodide. When benzalkonium was used, a 71 percent decomposition of 2,3,7,8-TCDD was observed. When cetylpyridinium chloroiodide was used, a 92 percent decomposition of 2,3,7,8-TCDD was achieved. Reaction products included chlorophenols, phenols, and 2-phenoxychloro-phenols. Quantitative information was not available for these substances. The results were obtained under ideal conditions, so extrapolations to actual decontamination should be made with great care.

Contaminated soil samples from Seveso, Italy were also treated (Botre, 1979; des Rosiers, 1983; Esposito, 1980). Samples were prepared by treating the soil with solutions containing surfactant micelles with chloroiodides and micelles without chloroiodides. A benzalkonium chloride micellar solution showed approximately a 14 percent decomposition of 2,3,7,8-TCDD. A solution containing benzalkonium chloroiodide in a micellar solution showed a decomposition of 52 percent of 2,3,7,8-TCDD. Thus, the addition of chloroiodides to micellar surfactant solutions greatly enhances the decomposition of 2,3,7,8-TCDD. This study did not specify whether or not exposure to UV radiation occurred. UV radiation may or may not enhance decomposition significantly, depending on the experimental configuration.

5.7.3 Costs of Treatment

Since no commercial applications exist for the use of chloroiodides to destroy TCDD, the economic feasibility cannot be estimated with any degree of confidence. Some cost factors of significance include:

1. the excavation and pretreatment of contaminated soil with a solubilizing solution;

2. washing of the solution from the soil;

3. extraction of 2,3,7,8-TCDD from the solution and concentration by evaporation; and

4. stringency of conditions needed to achieve a 1 ppb level with chloroiodide extracts.

5.7.4 Process Status

As stated earlier, the potential for the use of micellar surfactant solutions (with a UV source) may apply in decontaminating surfaces of buildings, furniture, and other personal belongings. The addition of 9 chloroiodides may improve the application of surfactant micellar solutions for this type of decontamination. The decomposition of 2,3,7,8-TCDD by chloroiodides (without a light source) has been proven in laboratory experiments. However, it has not been demonstrated that TCDD levels can be lowered to less than 1 ppb in soils. Additional bench scale testing is needed for further optimization of processes, perhaps including the possibility of in-situ decontamination of contaminated soil. In situ decontamination using solubilizing agents may not be feasible because it raises the possibility of causing the transport of 2,3,7,8-TCDD from soils into ground water and surface water. Chloroiodides have not destroyed all TCDD in clean liquid solutions, and it is unlikely that near 100% destruction in contaminated soils could be achieved.

5.8   GAMMA RAY RADIOLYSIS

5.8.1   Process Description

Gamma rays are electromagnetic waves of energy (photons) similar to x-rays, except that they are commonly generated in different ways and are of much higher energy. In fact, gamma rays possess the highest energy levels of all radiation in the electromagnetic spectrum. Gamma rays are emitted from the nucleus of radioactive substances as a result of transitions of protons and neutrons between two energy levels of the nucleus. X-rays, on the other hand, are the result of the de-excitation of electrons to a lower energy state. The energy of gamma rays ranges between 10 thousand electron volts (KeV) and 10 million electron volts (MeV).

The mechanism of gamma ray interaction with matter is a complex function of the radiation energy and the atomic number of the material (Kircher, 1964). At low energies, the gamma photon is completely absorbed by an electron and the electron is ejected from the atom (photoelectric effect). At higher energies, the photon can eject more strongly bound electrons, with the photon being scattered at a reduced energy (Compton effect). The scattered photon can also interact with electrons. At still higher energies, the gamma ray can interact with a nucleus and be absorbed, resulting in the production of two particles, a positive and a negative electron (pair production). The minimum photon energy for pair production is 1.02 MeV.

In each of the mechanisms described above, energetic electrons are produced, and it is this internal electron bombardment that actually causes chemical changes in a material irradiated by gamma rays. Rupture of chemical bonds results from the electron bombardment; thus organic hydrocarbons can be effectively degraded by gamma radiation.

Commercial sources of gamma radiation generally are unstable isotopes of cesium and cobalt. Gamma rays from cesium and cobalt-60 sources have energies in the range of 0.40 to 1.33 MeV. In this range, the primary mechanism of gamma ray interaction with matter is the Compton effect.

5.8.2   Technology Performance Evaluation

Gamma ray radiolysis is used on a commercial basis for many purposes, including thickness measurements in process control and for the sterilization of disposable medical products. The use of gamma rays for the degradation of chlorinated hydrocarbon wastes appears to be limited to investigations in research laboratories where the wastes were dissolved in various solvents (Fanelli, 1978; Buser, 1976; Craft, 1975). In each of the laboratory investigations, the major degradation pathway appears to be the dechlorination of the compounds to lower chlorinated compounds.

The effect of gamma rays on 2,3,7,8-TCDD dissolved in organic solvents was investigated in a series of preliminary experiments in Italy in 1978 (Fanelli, 1978). The investigators dissolved 2,3,7,8-TCDD in either ethanol, acetone, or dioxane at a concentration of 100 ng/ml (ppb). Irradiations of 0.5 ml samples were accomplished with a 10,000 Curie Cobalt-60 source. The dose rate was 106 rad/hour. It was found that the disappearance of 2,3,7,8-TCDD is directly related to the total dose of radiation absorbed and to the solvents used. About 97 percent of the 2,3,7,8-TCDD was degraded after 30 hours when ethanol was the solvent. Thus, the concentration of 2,3,7,8-TCDD was reduced from 100 ppb to about 3 ppb. Degradations of 80 and 70 percent, respectively, were achieved in 30 hours when acetone and dioxane were the solvents, as shown in Figure 5.8.1.

From the experiments described above, it seems clear that the type of solvent used is important for the efficiency of the degradation process. There was no irradiation testing of 2,3,7,8-TCDD in contaminated soil samples without solvents, although the authors indicated in their 1978 paper that further studies were in progress to verify the possible application of gamma rays to the degradation of 2,3,7,8-TCDD in contaminated soil samples.

In earlier related work in 1976, gamma ray experiments with octachlorodibenzo-p-dioxin and octachlorodibenzofuran dissolved in benzene and hexane at a concentration of 25 mg/L (ppm) were conducted (Buser, 1976). After 4 hours of gamma irradiation, 80 percent of the octachlorodibenzo-p-dioxin was converted to dioxin molecules containing 5, 6, or 7 chlorine atoms per molecule.

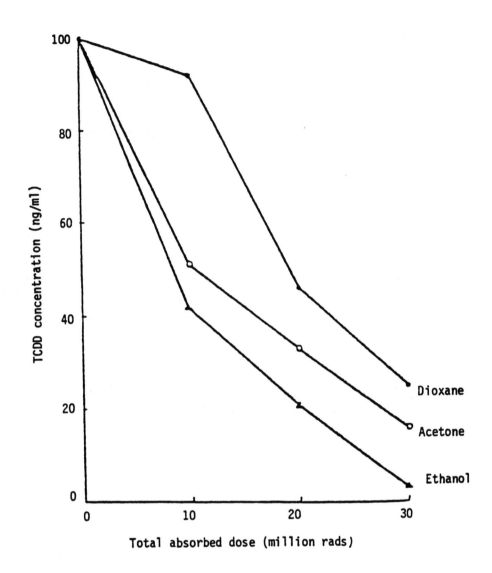

Figure 5.8.1. Effect of Gamma Ray irradiation on 2,3,7,8-TCDD
concentration in ethanol, acetone and dioxane
[Fanelli, 1978].

The potential for destroying pesticides using gamma ray radiation was investigated in 1975 at the Georgia Institute of Technology (Craft, 1975). Significant destruction of compounds such as pentachlorophenol, 2,4,5-T, and 2,4,-D were obtained, but no change in PCBs or mixtures of compounds such as "Herbicide Orange" could be detected. These researchers concluded that, because of the inefficiency of radiation in destroying mixtures of pesticides and dioxin molecules, radiation treatment of chlorinated hydrocarbons is not economically feasible.

5.8.3 Costs of Treatment

A prototype gamma ray radiolysis facility has been built at the Sandia National Laboratory. The facility is capable of treating about 4 tons per day of digested and dewatered sewage sludge at a dosage of 1 million rad. Sludge is passed by the radiation source in a bucket conveyor (1 to 1.5 ft^3 buckets). The capital cost of the facility is about $3 million and the processing cost is about $50 per ton (Pierce, 1984). For larger facilities (50 to 60 TPD), the processing cost could conceivably be reduced to as low as $10 per ton (Pierce, 1984).

5.8.4 Process Status

In summary, the development of gamma ray technology for the destruction of 2,3,7,8-TCDD is in the research stage. Based on some limited data, it appears that dechlorination occurs most readily when dioxin molecules are dissolved in certain solvents. If this technique were used for large quantities of contaminated soil, some method of removing the 2,3,7,8-TCDD from the soil (e.g., solvent extraction) would probably be required before gamma ray irradiation.

If further studies show that direct irradiation of 2,3,7,8-TCDD-contaminated soils also dechlorinates the dioxin molecule, then it may be possible to remove the contaminated soil and treat it at an irradiation facility, before replacement at the site. Because of the heavy radiation shielding required, it is not practical to treat large quantities of soil with portable units (Craft, 1984).

Further research is needed to verify the possible application of gamma ray radiolysis in the destruction of dioxin molecules in soils. Based on experimental data, it appears that a minimum dosage of 30 million rad of radiation is required to reduce the 2,3,7,8-TCDD level from 100 ppb to 30 ppb. (Fanelli, 1978).

REFERENCES

Acurex, Cincinnati, Ohio. Communication with representative. 17 May 1984.

Arisman, R.K., and R.C. Musik (General Electric Company); J.D. Zeftack,
    T.C. Crose (Westgate Research Corporation). Experience in Operation of a
    UV-Ozone (ULTROX) Pilot Plant for Destroying PCBs in Industrial Waste
    Effluent. Presented at the 35th Annual Purdue Industrial Waste
    Conference. May 1980.

Ayres, D. C., C. M. Scott. Oxidative Control of Organosulfur Pollutants.
    Environmental Science & Technology, 13(11):1383. November 1979.

Ayres, D. C. Destruction of Polychlorodibenzo-p-dioxins. Nature, 290:323.
    26 March 1981a.

Ayres, D. C. Ruthenium Tetroxide Destroys Dioxin. The Oxidative Control of
    Aromatic Pollutants. Platinum Met. Rev., 25(4):160. 1981b.

Ayres, D. C., D. P. Levy and C. S. Creaser. Destruction of Chlorinated
    Dioxins and Related Compounds by Ruthenium Tetroxide. In: Chlorinated
    Dioxins and Dibenzofurans in the Total Environment II. Butterworth
    Publishers, Stoneham, MA. 1985.

Banerjee, S., S. Duttagupta, A. M. Chakrabarty. Production of Emulsifying
    Agent During Growth of Pseudomonas Cepacia with 2,4,5-Tri-
    chlorophenoxyacetic Acid. Arch. Microbiology, 135:110-114. 1983.

Bartleson, F. D., Jr., D. D. Hanison, and J. B. Morgan. Field Studies of
    Wildlife Exposed to TCDD-Contaminated Soils. U.S. Air Force Armament
    Laboratory, Eglin Air Force Base, Florida. 1975.

Berry, R. I. New Ways to Destroy PCBs. Chemical Engineering, 88(16):37-41.
    August 10, 1981.

Bolton, L. Seveso Dioxin: No Solution in Sight. Chemical Engineering,
    85(22):78. 1978.

Botre, C. et al. 2,3,7,8-TCDD Solubilization and Photodegradation in Aqueous
    Solutions. Environmental Science and Technology. 12(3):335-336.
    March 1978.

Botre, C., Memoli, A., and Alhaique, F. Environmental Science and Technology,
    13(2):228. 1979.

Bumpus, J.A., Tien, M., Wright, D.A. and Aust, S.D. Biodegradation
    of Environmental Pollutants by the White Rot Fungus Phanerocheate
    chrysosporium. Presented at EPA HWERL 11th Annual Research Symposium,
    Cincinnati, OH, April 29-May 1, 1985.

Bumpus, J.A., Tien, M., Wright, D.A. and Aust, S.D.   Oxidation of
    Persistent Environmental Pollutants by a White Rot Fungus.   Science,
    228:1434-1436.   1985.

Buser, Hans-Rudolph.   Preparation of Qualitative Standard Mixtures of
    Polychlorinated Dibenzo-p-dioxins and Dibenzofurans by Ultraviolet and
    -Irradiation of the Octachloro Compounds.   Journals of Chromatography.
    129:303-307.   1976.

Centofanti, L. PPM, Inc. Personal Communication.   1986.

Camoni, I., et al.   Laboratory Investigation for the Microbial Degradation of
    2,3,7,8-tetrachlorodibenzo-p-dioxin in Soil by Addition of Organic
    Compost.   In: Chlorinated Dioxins and Related Compounds: Impact on the
    Environment. O. Hutzinger, et al., editors. Pergammon Press, New York,
    New York. pp. 95-103. 1982.

Chakrabarty, A. M. University of Illinois Medical Center at Chicago.
    Personal communication with M. Sutton, GCA Technology Division, Inc..
    17 January 1984.

Chatterjee, D. K., A. M. Chakrabarty.   Generic Rearrangements in Plasmids
    Specifying Total Degradation of Chlorinated Benzoic Acids. Mol. Gen.
    Genet, 188:279-285. 1982.

Chatterjee, D. K., A. M. Chakrabarty.   Genetic Homology Between Independently
    Isolated Chlorobenzoate-Degraditive Plasmids.   Journal of Bacteriology,
    153(1):532. January 1983.

Chemical Engineering. Commercial Scale Ultraviolet Destruction of Dioxin.
    88(18):18. September 7, 1981.

Commoner, B., and R. E. Scott.   Center for the Biology of Natural Systems,
    Washington University, St. Louis, Missouri. USAF Studies on the
    Stability and Ecological Effects of TCDD (Dioxin): An Evaluation
    Relative to the Accidental Dissemination of 2,3,7,8-TCDD at Seveso,
    Italy.   1976.

Corwell, D., Yamasaki, R. S. Journal of Chemical Physics, 2(5):1064-1065.
    November 1957.

Cotter, J. L., et al. TRW, Inc., Redondo Beach, CA. Facilities Evaluation of
    High Efficiency Boiler Destruction PCB Waste - Research Brief. January
    to April 1980. EPA Reprot No. EPA-600/7-81-031. NTIS. 1981.

Craft, T. F., R. D. Kimbrough, and C. T. Brown. Georgia Institute of
    Technology. Radiation Treatment of High Strength Chlorinated Hydrocarbon
    Wastes. U.S. EPA Report EPA-660/2-75-017. July 1975.

Craft, T.F. Georgia Institute of Technology. Telephone communication.
    May 1984.

Crosby, D.G., and A.S. Wong. Photodecomposition of Chlorinated
    Dibenzo-p-dioxins. Science. 173:748-749. August 20, 1971.

Crosby, D.G. Conquering the Monster: The Photochemical Destruction of
    Chlorodioxins. In: Disposal and Decontamination of Pesticides. ACS
    Symposium Series 73. Washington, DC. 1978.

Davis, L. Sybron Corporation. Personal communication to B. Farino,
    GCA Technology Division, Inc. 9 January 1984.

des Rosiers, P. Remedial Measures for Wastes Containing Polychlorinated
    Dibenzo-p-dioxins (PCDDs) and Dibenzo-furans (PCDFs): Destruction,
    Containment or Process Modification. Annals of Occupational Hygiene,
    27(1):57-72. 1983.

des Rosiers, P. Environmental Engineering and Technology, Office of Research
    and Development, EPA. Communication. 21 May 1984.

DiDominico, A., et al. Accidental Release of 2,3,7,8-tetrachlorodibenzo-p-
    dioxin (TCDD) at Seveso, Italy. Ecotoxicology and Environmental Safety,
    4(3):282-356. 1980.

Eaton, H. C., M. E. Tittlebaum, and F. K. Cartledge. Louisiana State
    University. Techniques for Microscopic Studies of Solidification
    Technologies. Proceedings of the 11th Annual Research Symposium on
    Incineration and Treatment of Hazardous Waste, sponsored by U.S.
    EPA-HWERL. Cincinnati, Ohio, April 29-May 1, 1985. EPA/600/9-85/028.
    September 1985.

Edwards, B.H., J.N. Paullin, and K. Coghlon-Jordon. Noyes Data Corporation.
    Emerging Technologies for the Control of Hazardous Wastes. pp. 76-87,
    105-108, 119-122. 1983.

Ellis, William D., William H. Vick, Donald E. Sanning, and Edward J. Opatkin.
    Evaluation of Stabilized Dioxin Contaminated Soils. Proceedings of the
    EPA-HWERL 11th Annual Research Symposium, Cincinnati, Ohio. April 29-May
    1, 1985.

Ellis, William. JRB Associates. Telephone Conversation with Lisa Farrell,
    GCA Technology Division, Inc. May 15, 1986.

Esposito, M. P., et al. Dioxins. EPA-600/2-80-197. 1980.

Esposito, M. P., et al. PEDCo Environmental, Inc. Dioxins: Vol. I, Sources,
    Exposure, Transport, and Control. Prepared for IERL. EPA-600/2-80-156.
    June 1980.

Esposito, M.P., T.O. Tiernon, and F.E. Dryden. Dioxins. EPA 600/2-80-197.
    pp. 263-264. November 1980.

Exner, J.H. et al. Process for Destroying Tetrachlorodibenzo-p-dioxin in a
    Hazardous Waste. Paper Presented in Detoxification of Hazardous Waste.
    Chapter 17. Ann Arbor Science, Ann Arbor, Michigan. 1982.

Fanelli, R., C. Chiabrando, M. Salmona, S. Garattini, and P. G. Calders. Degradation of 2,3,7,8-Tetrachlorodibenzo-p-dioxin in Organic Solvents by Gamma Ray Irradiation. Experentia. 34(9):1126-7. September 9, 1978.

Firestone, F. Oil and Hazardous Materials Spills Branch. Telephone conversation with T. Murphy, GCA/Technology Division. 19 January 1984.

Fisher, Marilee. SunOhio, Inc. Personal Communication. 1986.

Flax, L. Lopat Enterprises, Inc. Telephone Conversation with Lisa Farrell, GCA Technology Division, Inc. May 19, 1986.

Furukawa, K., A. M. Chakrabarty. Involvement of Plasmids in Total Degradation of Chlorinated Biphenyls. Applied and Environmental Microbiology, 44(3):619. September 1982.

Ghosal, D., et al. Microbial Degradation of Halogenated Compounds. Science, 220: 135-228.

Gibson. Personal communication with M. Sutton, GCA Technology Division, Inc. 17 January 1984.

Gilman, W. S. United States Testing Company, Inc., Chemical Services Division. Report of Test, January 4, 1983, February 7, 1983, February 17, 1983, May 31, 1983, June 13, 1983, August 4, 1983, August 19, 1983, August 23, 1983, October 13, 1983, October 21, 1983, December 22, 1983.

Goldensohn, R. Red Tape Slows Jersey Inventors' PCB eater. Sunday Star-Ledger, Neward, NJ, 70(223):68. 9 October 1983.

Goldsmith, D. Environmental Engineer, Sybirn Corporation. Personal Communication with M. Arienti, GCA Technology Division. April 29,1986.

Hay, A. Disposing of Dioxins by Oxidation. Nature, 290:294. March 26, 1981.

Hazardous Waste Consultant. Stabilizing Organic Wastes: How Predictable Are The Results? Volume 3, Issue 3, Pages 1-18 to 1-19. May/June 1985.

Helsel, R., et al. Technology Demonstration of a Thermal Desorption/UV Photolysis Process for Decontaminating Soils Containing Herbicide Orange. Preprint extended abstract. Presented before the Division of Environmental Chemistry. American Chemical Society. New York. April 1986.

Hutter, R., and M. Philippi. Studies on Microbial Metabolism of TCDD Under Laboratory Conditions. In: Chlorinated Dioxins and Related Compounds: Impact on the Environment. O. Hutzinger, et al., editors. Pergammon Press, New York, New York. pp. 87-93. 1983.

IT Corporation. Interim Summary Report on Evaluation of Soils Washing and Incineration as On-Site Treatment Systems for Dioxin-Contaminated Materials. EPA Contract No. 68-03-3069. 1985.

Jackson, N. E. SunOhio. Paper presented to the Annual Conference of the Southeastern Electric Exchange. April 1981.

Jiranders, J. Lopat Enterprises, Inc., Asbury Park, New Jersy. Telephone conversation with T. Murphy, GCA/Technology Division. 8 May 1984.

Karns, J. S., S. Duttagupta, A. M. Chakrabarty. Regulation of 2,4,5-Trichloro-phenoxyacetic Acid and Chlorophenol Metabolism in Pseudomonas Cepacia AC1100. Applied and Environmental Microbiology, 46(5):1182. November 1983a.

Karns, J. S., J. J. Kilbane, S. Duttagupta, A. M. Chakrabarty. Metabolism of Halophenols by 2,4,5-Trichlorophenoxyacetic Acid--Degrading Pseudomonas Cepacia. Applied and Environmental Microbiology, 46(5):1176. November 1983b.

Kearney, P. C., et al. Persistence and Metabolism of Chlorodioxins in Soils. Environmental Science and Technology. 6(12):1017-1019. 1972.

Kearney, P. C. Agricultural Research Center, Beltsville, Maryland. Personal communication with M. Sutton, GCA Technology Division, Inc.. 18 January 1984.

Kellogg, S. T., D. K. Chatterjee, A. M. Chakrabarty. Plasmid--Assisted Molecular Breeding: New Technique for Enhanced Biodegradation of Persistent Toxic Chemicals. Science, 214:1133-1135. December 1981.

Kilbane, J. J., D. K. Chatterjee, J. S. Karns, S.T. Kellogg, A. M. Chakrabarty. Biodegradation of 2,4,5-Trichlorophenoxyacetic Acid by a Pure Culture of Pseudomonas Cepacia. Applied and Environmental Microbiology, 44:72. June 1982.

Kilbane, J. J., D. K. Chatterjee, A. M. Chakrabarty. Detoxification of 2,4,5-Trichlorophenoxyacetic Acid from Contaminated Soil by Pseudomonas Cepacia. Applied and Environmental Microbiology, 45:1697. May 1983.

Kircher, J. F., and R. E. Bowman. Effects of Radiation on Materials and Components. Reinhold Publishing Corp., New York. 1964.

Kitchens, Judith. Atlantic Research Corporation. Personal Communication. 1986.

Klecka, G. M., and D. T. Gibson. Metabolism of Dibenzo-p-dioxin and Chlorinated Dibenzo-p-dioxins by a Beijerinckia Species. Appl. and Env. Microbiology, 39(2):288-296, 1980.

Klecka, G. M., D. T. Gibson. Bacterial Degradation of Dibenzo-p-dioxin and Chlorinated Dibenzo-p-dioxins. Prepared for Environmental Research Lab. EPA-600/4-81-016. March 1981.

Klee, Albert, et al. Report on the Feasibility of APEG Detoxification of Dioxin-Contaminated Soils. U.S. EPA-IERL/ORD. EPA-600/2-84-071. March 1984.

Legan, R.W. Ultraviolet Light Takes on CPI Role. Chemical Engineering,
    9(2):95-100. January 25, 1982.

Lubowitz, H. R., et al. Contaminant Fixation: Practice and Theory. Land
    Disposal of Hazardous Waste. Proceedings of the Tenth Annual Research
    Symposium. EPA-600/9-84-007. April 1984.

Malone, P. U.S. Army Engineer Waterways Experiment Station, Vicksburg,
    Mississippi. Telephone conversation with R. Bell, GCA/Technology
    Division. 27 March 1984.

McDaniel, P. Inventors Say They Can Neutralize Dioxin. Asbury Park Press,
    Asbury Park, NJ. p. A4. 12 June 1983.

Miille, G. J. Acurex Corporation. Paper presented to the PCB Seminar,
    sponsored by the Electric Power Research Institute. December 1981.

M. M. Dillon, Ltd. Destruction Technologies for Polychlorinated Biphenyls
    (PCBs). Prepared for Environment Canada, Waste Management Branch. 1982.

Peterson, R. L., et al. Chemical Destruction/Detoxification of Chlorinated
    Dioxins in Soils. Paper presented at Eleventh Annual Research Symposium
    on Incineration and Treatment of Hazardous Waste. EPA-600/9-85-028.
    September 1985.

Peterson, R. L., et al. Comparison of Laboratory and Field Test Data in the
    Chemical Decontamination of Dioxin Contaminated Soils Using the Galson
    PKS Process. Preprinted Extended Abstract. Presented before the
    Division of Environmental Chemistry. American Chemical Society.
    New York. April 1986.

Peterson, R.L. Galson Research Corporation, E. Syracuse, NY Telephone
    Conversation with M. Jasinski, GCA Technology Division, Inc. June 1986a.

Philippi, Martin, et al. Fate of 2,3,7,8-TCDD in Microbial Cultures and in
    Soil Under Laboratory Conditions. Fems Symp., Vol. 12, Iss. Microbial
    Degradation Xenobiotics Recalcitrant Compd. pp. 221-3. 1981.

Philippi, M., et al. A Microbial Metabolite of TCDD. Experientia,
    38:654-661. 1982.

Pierce, F. Sandia National Lab. Telephone Communication. May 1984.

Pocchiari, F. 2,3,7,8-Tetrachlorodibenzo-p-dioxin Decontamination. In:
    Chlorinated Phenoxy Acids and Their Dioxins. Ramel, editor. Ecol.
    Bulletin (Stockholm), 27:67-70. 1978.

Poiger, H., et al. Special Aspects of Metabolism and Kinetics of 2,3,7,8-TCDD
    in Dogs and Rats - Assessment of Toxicity of 2,3,7,8-TCDD Metabolites In
    Guinea Pigs. In: Chlorinated Dioxins and Related Compounds: Impact on
    the Environment. O. Hutzinger, et al., editors. Pergammon Press, New
    York, New York. pp. 317-324. 1983.

Quensen, J. F., and F. Matsumura. Oxidative Degradation of 2,3,7,8-tetra-
   chlorodibenzo-p-dioxin by Microorganisms. Environmental Toxicology.
   2(3):261-268. 1983.

Quensen, J. F. Dept. of Crop and Soil Sciences, Michigan State University.
   Personal Communication with M. Arienti, GCA Technology Division, Inc.
   April 29, 1986.

Quensen, J. Pesticide Research Center, East Lansing, Michigan. Personal
   communication with M. Sutton, GCA Technology Division, Inc.
   17 January 1984.

Rogers, C. J. Chemical Treatment of PCBs in the Environment. Paper presented
   at 8th Annual Research Symposium on Incineration and Treatment of
   Hazardous Waste. EPA-600/9-83-003. April 1983.

Rogers, C. J., et al. Interim Report on the Feasibility of Using UV
   Photolysis and APEG Reagent for Treatment of Dioxin Contaminated Soils.
   Project Summary. EPA-600/S2-85/083. December 1985.

Salkinoja-Sabnen, M. University of Helsinki, Finland. Microbial Dechlorination of Chloro-Organics. Presentation given at U.S. EPA Offices, IERL,
   Cincinnati, Ohio. December 1979.

Sawyer, C.J. Environmental Health and Safety Considerations for a Dioxin
   Detoxification Process. In: Detoxification of Hazardous Waste.
   Chapter 18. 1982.

Sferra, P. U.S. EPA Hazardous Waste Engineering Research Laboratory.
   Personal Communication with M. Arienti, GCA Technology Division, Inc.
   April 29, 1986.

Smith, Robert L., David T. Musser, Thomas J. DeGrood. ENRECO, Inc. In-Situ
   Solidification/Fixation of Industrial Wastes. Proceedings of the 6th
   National Conference on Management of Uncontrolled Hazardous Waste Sites,
   Washington, D.C. November 4-6, 1985.

Spooner, Philip A. Science Applications International Corporation (SAIC).
   Stabilization/Solidification Alternatives for Remedial Action.

SunOhio, Canton, Ohio. Marketing Brochure. 1985.

SunOhio, Canton, Ohio. Communication with representative. 17 May 1984.

Swarzgn, E.M., and D.G. Ackerman. Interim Guidelines for the Disposal/
   Destruction of PCBs and PCB Items by Nonthermal Methods.
   EPA-600/2-82-069. April 1982.

Technical Resources, Inc. Analysis of Technical Information to
   Support RCRA Rules for Dioxins-containing Waste Streams. Final Draft
   Report submitted to Paul E. des Rosiers, Chairman, U.S. EPA - Dioxin
   Advisory Group. July 31, 1985.

Telles, R. W., et al., Review of Fixation Processes to Manage Hazardous Organic Waste. Draft Report. Carlton Wiles, Project Officer, MERL, Cincinnati, Ohio. April 1984.

Tieran, T. O., et al. Dioxins. Industrial Environmental Research Laboratory, Office of Research and Development, Cincinnati, Ohio. EPA-600/2-80-197. November 1980.

Tieran, T. O. Chlorodibenzodioxins and Chlorodibenzofurans: An Overview, Detoxification of Hazardous Waste. Ann Arbor Science, Ann Arbor, Michigan. p. 245. 1982.

Tittlebaum, Marty E., et al. State-of-the-Art on Stabilization of Hazardous Organic Liquid Wastes and Sludges. In: Critical Reviews in Environmental Control, 15(2):179-211. 1985.

Tridje, J. Michigan State University. Personal communication with M. Sutton, GCA/Technology Division. 18 January 1984.

Tundo, P. Chemical Degradation of 2,3,7,8,-TCDD By Means of Polyethylene-glycols in the Presence of Weak Bases and an Oxidant. In: Chemosphere. Volume 14, No. 5, pp. 403-410. 1985.

Tumiatti, W. Site Decontamination and Chemical Degradation of PCDFs and PCDDs Coming From Pyrolysis of PCBs. Preprinted Extended Abstract. Presented Before the Division of Environmental Chemistry. American Chemical Society. New York. April 1986.

U.S. Army Engineers, Waterways Experiment Station, Environmental Laboratory, Vicksburg, Mississippi. Guide to the Disposal of Chemically Stabilized and Solidified Waste. Report prepared for U.S. EPA, Solid and Hazardous Waste Research Division, Municipal Environmental Research Laboratory, Cincinnati, Ohio. EPA-IAG-D4-0569. September 1982.

Valentine, R.S. LARC-Light Activated Reduction of Chemicals. Pollution Engineering. 1981.

Vanness, G. F., et al. Tetrachlorodibenzo-p-Dioxins in Chemical Wastes, Aqueous Effluents and Soils. Chemosphere. 9(9):553-63. 1980.

Vick, W.H., S. Denzer, W. Ellis, J. Lambauch, and N. Rottunda. Evaluation of Physical Stabilization Techniques for Mitigation of Environmental Pollution from Dioxin-Contaminated Soils. Interim Report: Summary of Progress-To-Date. Submitted to EPA-HWERL by SAIC/JRB Associates, EPA Contract No. 68-03-3113, Work Assignment No. 36. June 1985.

Ward, C. T., et al. Fate of 2,3,7,8-Tetrachlorodibenzo-p-Dioxin (2,3,7,8-TCDD) in a Model Aquatic Environment. Archives of Environmental Contamination and Toxicology. 7:349-357. 1978.

Waste Age. Destroying Dioxin: A Unique Approach. pp. 60–63. October 1980.

Weitzman, L. Acurex Corporation, Cincinnati, Ohio. Telephone conversation
    with M. Jasinski, GCA Technology Division, Inc. 4 June 1986.

Weitzman, L. Treatment and Destruction of Polychlorinated Biphenyls and
    Polychlorinated Biphenyl-Contaminated Materials. In: Detoxification of
    Hazardous Waste, Chapter 8. 1982.

Weitzman, L. Acurex Waste Technologies, Inc. Telephone conversation with
    T. Murphy, GCA Technology Division, Inc. 19 January 1984.

Worne Biotechnology, Medford, New Jersey. Personal communication with
    Dr. Worne. 1984.

Young, A. L., et al. Fate of 2,3,7,8-Tetrachlorodibenzo-p-dioxin
    (TCDD) in the Environment: Summary of Decontamination Recommendations.
    USA FA-TR-76-18. 1976.

Young, A. L. Long-Term Studies on the Persistence and Movement of TCDD in a
    Natural Ecosystem: In: Human and Environmental Risks of Chlorinated
    Dioxins and Related Compounds. R. E. Tucker, et al., editors. Plenum
    Press, New York, New York. pp. 173–190. 1983.

Zepp, R.G., and D.M. Kline. Rates of Direct Photolysis in Aquatic Environment.
    Environmental Science and Technology. 11(4):359–366. April 1977.

6. Factors Affecting Technology Selection

Section 3 contained information on the quantities and types of dioxin wastes generated by industrial processes and residuals identified as sources of dioxin-containing waste. Table 6.1 summarizes information on their sources, their usual physical form, and estimates of present and future quantities of wastes generated within each EPA waste code. Sections 4 and 5 contained information on the technical aspects of a number of potential treatment technologies for these wastes. This information is summarized in Table 6.2.

The purpose of this section is to review this previously developed information and identify factors which would affect the selection/use of a particular technology for treating a specific waste type. This document has been concerned largely with the assessment of the technical factors relating to treatment technology performance. However, both technical performance and cost will generally be considered when selecting the most appropriate process for a specific waste stream. Both are considered in the following discussions.

6.1 TECHNICAL REQUIREMENTS FOR PROCESS SELECTION

Key factors which should be considered in assessing the technical applicability of treatment technologies to specific waste streams include:

1. Has the technology demonstrated that it can achieve 99.9999 percent DRE on CDD (or similar compounds)?

TABLE 6.1. SUMMARY OF DIOXIN WASTE SOURCES AND QUANTITIES

Waste code	Waste source	Physical form	Quantity generated (metric tons)	
			Present (or stored)	Future
F020	Manufacture of herbicides such as 2,4,5-T, 2,4,5-trichlorophenol, hexachlorophene; disposal of wastes in uncontrolled landfills or storage areas	- Still bottoms containing organic solvents and chloro-phenols - Nonaqueous phase leachate (NAPL) containing solvents, chlorophenols, heavy metals - Carbon used to treat aqueous leachate	Still bottoms - 2,300 NAPL - 1,450 Other - 550	0 0 - 200 Unknown
F021	Manufacture of pentachlorophenol: wastes from purification; wastes from formulation	- Still bottoms or other concentrated materials containing nonvolatile organic solids and chlorinated solvents and phenols - Sludges from formulation	Still bottoms - 0 Formulation waste-700	750' Unknown
F022	No known sources at this time	- NA*	0	0
F023	Production of chemicals on equipment formerly used to manufacture F020 compounds, e.g., 2,4-D on 2,4,5-T equipment	- Similar to F020 wastes - still bottoms, reactor residues containing chloro-phenols and organic solvents, and wash water sludges from formulation	0 - 600	0 - 600
F026	No known sources at this time	- NA	0	0
F027	Discarded formulation of tri-, tetra-, and pentachlorophenols and their derivatives	- Active ingredient in an emulsifiable concentrate, as a salt or an ester, or dissolved in an oil (such as in the case of pentachloro-phenol	1000 - 2000	0-1,000***
	Contaminated soil from improper disposal and spills of F020-F027**	- Soils containing low concentrations of dioxins and related compounds	500,000	Unknown

*NA - Not available
**Not listed as a specific waste code
***Only from pentachlorophenol products

TABLE 6.2. SUMMARY OF TREATMENT PROCESSES

Process name	Applicable waste streams	Stage of development	Performance/destruction achieved	Cost	Residuals generated
Stationary Rotary Kiln Incineration	Solids, liquids, sludges	Several approved and commercially available units for PCBs; not yet used for dioxins	Greater than six nines DRE for PCBs; greater than five nines DRE demonstrated on dioxin at combustion research facility	$0.25 - $0.70/lb for PCB solids	Treated waste material (ash), scrubber wastewater, particulate from air filters, gaseous products of combustion
Mobile Rotary Kiln Incineration	Solids, liquids, sludges	EPA mobile unit is permitted to treat dioxin wastes; ENSCO unit has been demonstrated on PCB waste	Greater than six nines DRE for dioxin by EPA unit; process residuals delisted	NA*	Same as above.
Liquid Injection Incineration	Liquids or sludges with viscosity less than 10,000 ssu (i.e., pumpable)	Full scale land-based units permitted for PCBs; only ocean incinerators have handled dioxin wastes	Greater than six nines DRE on PCB wastes; ocean incinerators only demonstrated three nines on dioxin containing herbicide orange	$200 - $500/ton	Same as above, but ash is usually minor because solid feeds are not treated
Fluidized-bed Incineration (Circulating Bed Combustor)	Solids, sludges	GA Technologies mobile circulating bed combustor has a TSCA permit to burn PCBs anywhere in the nation; not tested yet on dioxin	Greater than six nines DRE demonstrated by GA unit on PCBs	$60 - $320/ton for GA unit	Treated waste (ash), particulates from air filters
High Temperature Fluid Wall (Huber AER)	Primarily for granular contaminated soils, but may also handle liquids	Huber stationary unit is permitted to do research on dioxin wastes; pilot scale mobile reactor has been tested at several locations on dioxin contaminated soils	Pilot scale mobile unit demonstrated greater than five nines DRE on TCDD - contaminated soil at Times Beach (79 ppb reduced to below detection)	$300 - $600/ton	Treated waste solids (converted to glass beads), particulates from baghouse, gaseous effluent (primarily nitrogen)
Infrared Incinerator (Shirco)	Contaminated soils/sludges	Pilot scale, portable unit tested on waste containing dioxin; full scale units have been used in other applications; not yet permitted for TCDD	Greater than six nines DRE on TCDD-contaminated soil	Treatment costs are $200 - $1,200 per ton	Treated material (ash); particulates captured by scrubber (separated from scrubber water)

*Not available

(continued)

TABLE 6.2 (continued)

Process name	Applicable waste streams	Stage of development	Performance/destruction achieved	Cost	Residuals generated
Molten Salt (Rockwell Unit)	Solids, liquids, sludges; high ash content wastes may be troublesome	Pilot scale unit was tested on various wastes - further development is not known	Up to eleven nines DRE on hexachlorobenzene; greater than six nines DRE on PCB using bench scale reactor	NA	Spent molten salt containing ash, particulates from baghouse
Supercritical Water Oxidation	Aqueous solutions or slurries with less than 20 percent organics can be handled	Pilot scale unit tested on dioxin-containing wastes - results not yet published	Six nines DRE on dioxin-containing waste reported by developer, but not presented in literature; lab testing showed greater than 99.99% conversion of organic chloride for wastes containing PCB	$0.32 - $2.00/gallon; $77 - $480/ton	High purity water, inorganic salts, carbon dioxide, nitrogen
Plasma Arc Pyrolysis	Liquid waste streams (possibly low viscosity sludges)	Prototype unit (same as full scale) currently being field tested	Greater than six nines destruction of PCBs and CCl_4	$300 - $1,400/ton	Exhaust gases (H_2 and CO) which are flared and scrubber water containing particulates
In Situ Vitrification	Contaminated soil - soil type is not expected to affect the process	Full scale on radioactive waste; pilot scale on organic contaminated wastes	Greater than 99.9% destruction efficiency (DE) (not offgas treatment system) on PCB-contaminated soil	$120 - $250/m³	Stable/immobile molten glass; volatile organic combustion products (collected and treated)
Solvent Extraction	Soils, still bottoms	Full scale still bottoms extraction has been tested - pilot scale soils washer needs further investigation	Still bottom extraction: 340 ppm TCDD reduced to 0.2 ppm; 60-90% removal from soils, but reduction to below 1 ppb not achieved	NA	Treated waste material (soil, organic liquid); solvent extract with concentrated TCDD
Stabilization/ Fixation	Contaminated soil	Laboratory scale using cement and emulsified asphalt; lab tests also using K-20	Tests using cement showed decreased leaching of TCDD, but up to 27% loss of stabilized material due to weathering followed by leaching	NA	Stabilized matrix (soil plus cement, asphalt, or other stabilization material); matrix will still contain TCDD

(continued)

TABLE 6.2 (continued)

Process name	Applicable waste streams	Stage of development	Performance/ destruction achieved	Cost	Residuals generated
UV Photolysis	Liquids, still bottoms, and soils can be treated if dioxin is first extracted or desorbed into liquid	Full scale solvent extraction/UV process was used to treat 4,300 gallons of still bottoms in 1980; thermal desorption/UV process currently undergoing second field test	Greater than 98.7% reduction of TCDD using solvent extraction/ UV process - residuals contained ppm concentrations of TCDD; thermal desorption/UV process demonstrated reduction of TCDD in soil to below 1 ppb	Cost of treating the 4,300 gallons of still bottoms using solvent extraction/UV was $1 million; thermal desorption/UV estimated to cost $250 - $1,250/ton	Solvent extraction/UV process generated treated still bottoms, a solvent extract stream, and an aqueous salt stream; thermal desorption/UV generates a treated soil stream and a solvent extract stream
Chemical Dechlorination- APEG processes	Contaminated soil (other variations of the process used to treat PCB-contaminated oils)	Slurry process currently being field tested at pilot scale; in situ process has been tested in the field	Laboratory research has demonstrated reduction of 2,000 ppb TCDD to below 1 ppb for slurry (batch process); laboratory and field testing of in situ process not as promising	$296/ton for in situ APEG process; $91/ton for slurry (batch) process	Treated soil containing chloride salts (reagent is recovered in the slurry process)
Biological Degradation- primarily in situ addition of microbes	Research has been directed toward in situ treatment of contaminated soils - liquids are also possible	Currently laboratory scale-field testing in next year or two	50-60% metabolism of 2,3,7,8- TCDD in a week long period under lab conditions using white rot fungus - reduction to below 1 ppb not achieved	NA	Treated waste medium such as soil or water with TCDD metabolites depending on microorganisms
Chemical Degradation using Ruthenium Tetroxide	Liquid or soil wastes - possible most effective in decontaminating furniture, other surfaces	Laboratory scale - no work reported since 1983	Reduction of 70 ppb TCDD to below 10 ppb in 1 hr (on soil sample)	NA	Treated medium plus the solvent which has been added (water, CCl_4); TCDD end products not known
Chemical Degradation using Chloroiodides	Liquid or soil - thought to be most applicable to decontaminating furniture and buildings	Laboratory scale - no work reported since 1983	Up to 92% degradation on solution of TCDD in benzene - reductions to below 1 ppb were not demonstrated	NA	Treated waste medium; degradation end products are chlorophenols
Gamma Ray Radiolysis	Liquid waste streams (has been applied to sewage sludge disinfection)	Laboratory research; no research currently being conducted	97% destruction of 2,3,7,8-TCDD in ethanol after 30 hours - 100 ppb to 3 ppb	Cost for sewage disinfection facility treating 4 tons per day is $40 per ton; TCDD treatment would be more expensive	Less chlorinated dioxin molecules are the degradation end products in addition to the treated waste medium

2. Has the technology demonstrated that residues contain less than
    1 ppb CDDs and CDFs?

3. Can the treatment unit or process be transported to the waste site?

4. Can the process be used on in situ wastes or must the waste material
    first be removed and then fed to the unit?

5. Can the treatment process handle wastes of all physical states or is
    it limited to just liquids or just solids (and to what extent is
    pretreatment possible and/or justified)?

6. Is the process a final treatment/destruction process or just a
    temporary, or pretreatment process?

Responses to the above considerations have been provided in matrix form
in Table 6.3, Treatment Technology Selection Chart, for each of the
technologies discussed in Sections 4 and 5. A discussion of this table and
its significance with regard to the assessment and selection of particular
technologies is provided below.

6.1.1 Demonstration of Six Nines DRE

The dioxin listing rule specifies that incinerators and other thermal
treatment units must achieve 99.9999 percent (six nines) destruction and
removal efficiency (DRE) for CDDs and CDFs in order to become fully
permitted. Three technologies have demonstrated this level of performance.
These are the EPA mobile (rotary kiln) incinerator, the Shirco infrared
incinerator, and the Huber Advanced Electric Reactor (AER). Achievement of
six nines DRE was obtained by Huber on tests of its small-scale stationary
research reactor located at its Borger, Texas facility. Due to the
concentration of TCDD in the waste feed, their pilot scale, transportable unit
was only able to demonstrate five nines DRE. Another technology that has
reportedly achieved six nines DRE on CDDs is the supercritical water oxidation
process. Data to indicate the exact conditions under which this was achieved
have not yet been released by the developer of the process.
    In addition to those processes that have demonstrated six nines DRE on
CDDs and CDFs, processes that have demonstrated six nines DRE on compounds
that are at least as difficult to destroy as CDDs and CDFs, such as PCBs, may

TABLE 6.3.   TREATMENT TECHNOLOGY SELECTION CHART

Process	Has demonstrated six nines DRE on dioxin and/or reduction of dioxin in residuals to below 1 ppb	Has demonstrated six nines DRE of PCBs	Mobile or transportable process can be constructed	Process can be carried out in situ (without excavation of soil)	Can treat solids such as soils and heavy sludges	Can treat liquids and low viscosity sludges	Currently being investigated with regard to dioxin waste	Is a pre-treatment or temporary process	Is a final process
Stationary rotary kiln		x			x[h]	x	x		x
Mobile rotary kiln	x[a]	x	x		x[h]	x	x		x
Liquid injection incinerator		x	x			x			x
Fluidised bed (circulating bed combustor)		x	x		x[h]	x			x
High temperature fluid wall (Huber AER)	x[b]	x	x		x[h]	x	x		x
Infrared incinerator (Shirco)	x		x		x	x	x		x
Plasma arc		x	x			x	x		x
Molten salt		x	x[e]		x[g]	x	x		x
Supercritical water	x[c]		x		x[i]	x[h]	x		x
In situ vitrification				x	x				x
Solvent extraction					x[h]	x	x	x	
Stabilization/fixation				x	x	x	x	x	
UV photolysis	x[c]				x[i]	x	x		x
Chemical dechlorination (APEG)	x[d]		x	x[f]	x[h]	x	x		x
Biodegradation				x	x[h]	x	x		x
Ruthenium tetroxide					x[h]	x			x
Chloroiodides					x	x[h]			x
Gamma ray radiolysis						x[h]			x

a. EPA mobile rotary kiln.
b. Their stationary unit is permitted to do research on dioxin wastes.
c. Developer has indicated this, but presented no data.
d. during laboratory scale equipment.
e. One developer is designing a mobile unit.
f. There exists both an in situ and a batch reactor process.
g. High ash wastes may pose problems.
h. Indicates primary waste type.
i. solids only treated if some sort of extraction/desorption process removes the dioxin from soil.

also have potential for use. In particular, those processes that have TSCA permits to treat PCBs, which include several rotary kiln and liquid injection incinerators, may be able to burn listed dioxin waste if their previous trial burn demonstrates the six nines performance standard on compounds such as PCBs (51 FR 1733).

Processes that have achieved six nines DRE on CDDs and CDFs, or on PCBs, have demonstrated that they can technically be considered as candidates for treating listed dioxin wastes. Other factors that will affect the selection of an appropriate technology are considered below.

6.1.2   Demonstration of Treatment Residuals with Less Than 1 ppb of CDDs/CDFs

The proposed land disposal restrictions specify that the residuals from treatment of listed dioxin wastes must contain less than 1 ppb of extractable CDDs (and CDFs) in order for them to be land disposed as nonhazardous materials. Not only must the gases exiting an incinerator or thermal treatment process be virtually free of CDDs and CDFs, but so also should be the treated ash, scrubber water, filter residues, and other residuals. In some cases, the amount of dioxin in the waste feed may not be great enough to allow demonstration of six nines DRE, however, the concentration of dioxin in the treated residues may still be reduced to below detection limits (1 ppb). Two examples of this occurrence are the test burns at the Combustion Research Facility and the test burns using the Huber transportable AER. In both of these cases, the concentration of dioxin in the residual streams was below detection limits, but six nines was not demonstrated because the analytical detection limits were not low enough. Nonetheless, the data seemed to indicate that the treatment processes were effective (Ross et al., 1986; Roy F. Weston, 1985).

In addition, for nonthermal treatment methods such as chemical dechlorination, destruction and removal efficiency (DRE) is not an applicable measure of performance. A more appropriate measure is the residual concentration of contaminants in the treated waste. Nonthermal treatment processes that can achieve less than 1 ppb of CDDs and CDFs in the treated waste have presumably demonstrated their potential for use on actual dioxin wastes.

6.1.3 Mobile/Transportable Technology

The ability to bring the waste treatment unit to the waste site is very important, particularly when treating "dioxin" waste. The transportation of dioxin waste is very controversial, and has been opposed by the public in several instances. For example, an attempt was made to obtain permission to transport dioxin-containing leachate to the SCA incinerator in Chicago. Illinois residents strongly objected to this, and local authorities indicated that drastic measures would be used to block the effort (Gianti, 1986). As a result, many of the developing thermal and nonthermal technologies are being designed to function as mobile/transportable units that can be taken to the waste site. The units that are mobile are identified in Table 6.3. Stationary treatment facilities, even though they are able to demonstrate high levels of destruction, may not be fully utilized for highly toxic dioxin wastes.

In addition to avoiding the risk of spillage during transportation, another advantage of the use of a mobile unit is that the cost of transporting the waste to the treatment facility is eliminated. For each of these reasons, processes that are designed to be mobile appear to be more useful for treating listed dioxin wastes.

6.1.4 In Situ Technology

Similarly, processes than can treat the waste in situ may also be advantageous. In situ processes are aimed primarily at contaminated soil. Most processes require that the soil be excavated and then be fed to the treatment process. A process in which the excavation step is eliminated may be more environmentally and economically acceptable than a process that relies on excavation of the waste. Not only is excavation expensive, but it may also result in the dispersal of contaminated soil particles and greater human exposure to the contaminant. Processes with the potential for in situ use include chemical dechlorination, in situ vitrification, biodegradation, and stabilization/fixation. Chemical dechlorination, using potassium polyethylene glycol (KPEG), has been tested both as an in situ and a batch reactor type of process. As indicated in Section 5, the batch reactor variation of the process has been much more successful to date, both in terms of destruction of dioxin, and in terms of cost. The estimated cost for the batch reactor process is less, primarily because the chemical reagent can be recovered when the process is conducted in the laboratory under enclosed and controllable conditions. However, when the reagent is added directly to the soil as it is in the field, it cannot be recovered. The cost savings associated with the offsite treatment approach offset the cost of excavation. In general, an in situ process is desirable whenever there is a large quantity of contaminated soil in which the level of contamination is not extremely high (10 to 100 ppb). In these cases the quantity of soil that would have to be excavated to destroy a small quantity of dioxin may not be justified.

6.1.5 Waste Physical Form

The physical form of a waste will have a large effect on the choice of an appropriate treatment technology. As indicated in Table 6.1, the largest quantity of dioxin-bearing waste falls into the category of contaminated soils. The second largest quantity falls into the category of organic liquids. These include still bottoms and other process wastes from the manufacture of 2,4,5-trichlorophenol and chlorophenoxy herbicides, and non aqueous phase leachate from the landfilling of waste from these manufacturing processes. The quantity of contaminated soil is estimated to be at least 500,000 metric tons (MT), while the maximum quantity of organic liquid waste currently awaiting treatment is estimated to be approximately 7,500 MT. Consequently, it would be expected that the development of technologies to treat contaminated soil is much more important than the development of technologies to treat organic liquids. Conversely, the organic liquids are generally contaminated with CDDs and CDFs to a much higher degree than are the contaminated soils. In addition, since they are liquids, their mobility, if released to the environment is higher. Therefore, it is important to develop technologies to treat both types of waste.

Several of the technologies are designed to be used specifically for the treatment of either solids or liquids. For solids, these include in situ vitrification, infrared incineration, and stabilization/fixation. For liquids, these include plasma arc pyrolysis, liquid injection incineration and gamma ray radiolysis. Other technologies, including both fluidized bed and rotary kiln incineration, chemical dechlorination, and the high temperature fluid wall process have been used to treat both liquid and solid wastes. Finally, other processes are designed to be used primarily for either liquids or solids, but if certain pretreatment measures are applied, they may be able to treat both waste forms. For example, ultraviolet (UV) photolysis is only effective in treating waste streams that the radiation can penetrate, such as a nonabsorbing liquid. If, however, the contaminant is first desorbed, either thermally or using an organic solvent, a contaminated soil waste can be treated using this technology.

Perhaps the most important factor with regard to the physical properties of a waste stream is that each waste stream must be treated individually, and variations in waste characteristics fully assessed, to ensure that unanticipated difficulties do not arise. Even though pilot or laboratory scale data may indicate that a certain waste form is easily treatable, processing of the actual waste stream may pose problems. One example of this is the trial burns of Vertac still bottoms at the Combustion Research facility. The waste stream, on the basis of preliminary evaluations, was originally pumped through a feed lance into the rotary kiln without dilution or mixing with fuel. However, the lance frequently became inoperative due to clogging and carbon buildup, and the waste feed had to be interrupted so the lance could be cleaned. The problem was finally rectified by diluting the waste with water prior to pumping it into the kiln (Ross et al., 1986). Another example is the application of KPEGs directly to soil (in situ) to dechlorinate TCDD. In the lab, the process was fairly successful, but when applied in the field the KPEG reagent was seriously degraded by moisture in the soil, and the resulting degradation of TCDD was minimal. A third example involves the Huber AER. This unit was originally tested on very granular, uniform materials such as sand. In reality, all contaminated solids or soils are not as dry and uniformly graded as sand. Consequently, when the Huber reactor was tested on actual waste, there were problems with the feed mechanism. Special pretreatment measures to produce a granular, free-flowing feed had to be incorporated (Boyd et al., 1986).

To summarize, waste characteristics and process capabilities must be carefully evaluated before the appropriate treatment technology can be applied.

6.1.6 Pretreatment Versus Final Treatment/Destruction

Two of the processes in Table 6.3 are not classified as final treatment processes; e.g., solvent extraction and stabilization/fixation. Solvent extraction is potentially a very attractive method of separating CDDs and CDFs from soil or other waste matrices so that a less voluminous waste stream containing a higher concentration of contaminant may undergo final treatment/destruction. Solvent extraction will always be followed by another process such as liquid injection incineration or photolysis as was done in the case of the 4300 gallons of still bottoms at the Syntex plant. Unfortunately, solvent extraction has not yet been demonstrated capable of removing CDDs to residual levels below 1 ppb.

Stabilization/fixation is generally a temporary process. It, like landfilling, contains the waste, but does not generally involve destruction or true chemical fixation of the contaminants. Processes such as this may be useful in cases where contamination or release rate levels are low. In the end, however, it should be recognized that a final treatment/destruction process, in which CDDs and CDFs are destroyed, may be required.

6.2 COST OF TECHNOLOGY

Another important factor in the selection of a treatment technology, although not listed in Table 6.3, is the cost of treating the waste. Many of the technologies discussed are innovative technologies for which accurate estimates of treatment costs have not yet been developed. In general, however, thermal technologies are the most expensive to use, particularly for certain dioxin wastes such as contaminated soils, since high energy input is required to treat small quantities of contaminants. In addition, thermal technologies generally require the use of expensive equipment to treat both the waste itself and the off gases and other residuals that result. However, one cannot generalize with respect to thermal technologies either, since there may be a wide range of costs depending on the physical form of the waste and the specific technology used. For example, rotary kiln incineration of contaminated soil may cost 800 to 900 dollars/ton, while liquid injection incineration of a halogen-containing liquid waste may cost 200 dollars/ton.

The cost of treating dioxin-containing waste is very much affected by the high level of risk associated with the treatment of these waste streams and their residuals. Generally, processing of these wastes requires the imposition of extraordinary and often redundant measures to ensure that risks are not incurred by workers and the general population. One example of this was the attempted incineration several years ago of 6,000 gallons of solvent waste contaminated with 14 ppb of TCDD. The Dioxin Disposal Advisory Group (DDAG) recommended incineration of the waste at the ENSCO incinerator in Arkansas. ENSCO usually charges $325 per 55-gallon drum of hazardous waste, but in this case they would have charged $45,000 for the waste plus a $150,000 surcharge because it contained dioxin (Technical Resources, Inc. 1985). This amounts to a unit cost of approximately $30/gallon ($1,500 per 55-gallon drum) or $6,000/ton for 10 lb/gallon waste. In the end, however, ENSCO refused to accept the waste because of overwhelming public opposition (and implied liability).

Cost will always be an important factor in the selection of a treatment method, but at this time the demonstration of technical and environmental effectiveness appears to a more overriding concern. Technology must be fully demonstrated for treating dioxin wastes if public concerns are to be addressed and reconciled with the need for effective treatment.

6.3 SUMMARY

Each of the technical and cost factors discussed above will affect the final selection of a technology to treat the waste. It is important to keep in mind that the field of dioxin waste treatment is in a developmental stage. At the present time, only a few of the technologies have demonstrated 6 nines DRE on CDDs and CDFs, although many of the technologies are now undergoing performance testing using dioxin waste, or will undergo testing in the near future. In addition, with the ban on land disposal of dioxin wastes scheduled to go into effect in November 1986, work on the development of additional new technologies for treating these wastes can be expected to accelerate. Information in this document represents the developmental status of dioxin waste treatment technologies in the spring of 1986; revisions will be required as anticipated technical advances are made in the future.

REFERENCES

Boyd, James. Huber Corporation. Telephone conversations with Lisa Farrell, GCA Technology Division, Inc. January 28, 1986; April 3, 1986; May 1, 1986.

Gianti, S. U.S. EPA Region II. Telephone conversation with M. Arienti, GCA Technology Division, Inc. March 6, 1986.

Ross, R.W., II, T.H. Backhouse, R.N. Vogue, J.W. Lee, and L.R. Waterland. Acurex Corporation, Energy & Environmental Division, Combustion Research Facility. Pilot-Scale Incineration Test Burn of TCDD-Contaminated Toluene Stillbottoms from Trichlorophenol Production from the Vertac Chemical Company. Prepared for the U.S. EPA, Office of Research and Development, Hazardous Waste Engineering Research Laboratory, under EPA Contract No. 68-03-3267, Work Assignment Nos. 0-2. Acurex Technical Report TR-86-100/EE. January 1986.

Roy F. Weston, Inc. and York Research Consultants. Times Beach, Missouri: Field Demonstration of the Destruction of Dioxin in Contaminated Soil Using the J.M. Huber Corporation Advanced Electric Reactor. February 11, 1985.

Technical Resources, Inc. Final Draft Report: Analysis of Technical Information to Support RCRA Rules for Dioxin-containing Waste Streams. Submitted to Paul E. des Rosiers, Chairman, U.S. EPA - Dioxin Advisory Group. July 31, 1985.

Other Noyes Publications

TREATMENT TECHNOLOGIES
FOR SOLVENT CONTAINING WASTES

by

**M. Breton, P. Frillici, S. Palmer
C. Spears, M. Arienti, M. Kravett
A. Shayer, N. Surprenant**
Alliance Technologies Corporation

Pollution Technology Review No. 149

This book provides technical information describing management options for solvent containing wastes. These options include treatment and disposal of waste streams as well as waste minimization procedures such as source reduction, reuse, and recycling.

Emphasis is placed on proven technologies such as incineration, use as fuel, distillation, steam stripping, biological treatment, and activated carbon adsorption; however, a full range of waste minimization processes and treatment recovery technologies which can be used to manage solvent wastes is covered in the book.

Potentially viable technologies are described in terms of performance in removal of regulated constituents, associated process residuals and emissions, and restrictive waste characteristics which affect the ability of a given technique to effectively treat the wastes under consideration.

Approaches to the selection of treatment/recovery options are reviewed, and pertinent properties of organic solvents which impact treatment technology/waste interactions are provided.

CONTENTS

ISBN 0-8155-1158-2 (1988)

753 pages

Other Noyes Publications

BIOTECHNOLOGY FOR DEGRADATION OF TOXIC CHEMICALS IN HAZARDOUS WASTES

Edited by

R.J. Scholze, Jr., E.D. Smith, J.T. Bandy
U.S. Army Construction Engineering Research Lab

Y.C. Wu
New Jersey Institute of Technology

J.V. Basilico
U.S. Environmental Protection Agency

The state-of-the-art of biotechnology for degradation of toxic chemicals in hazardous wastes is discussed in this book. In particular, it discusses the applicability of using biotechnology for the treatment of hazardous/toxic wastewaters.

Full-scale application of biotechnology for the treatment of municipal and industrial wastewaters has been practiced for many years. However, whether this technology can be employed for detoxification and destruction of hazardous chemicals in aqueous and solid media is not yet fully understood. Removal of toxic and refractory organics in wastewater, groundwater, and leachate may be more efficient as a result of combining biological treatment with other treatment technologies such as chemical and physical methods. Development of standard techniques for biotoxicity detection and toxicity reduction evaluation is essential and extremely important to both technical determination and decisions on the future policy for hazardous waste management. The various chapters in the book describe current research in biotechnology for degradation of toxic chemicals in hazardous wastes.

A condensed table of contents containing **selected chapter titles** is given below.

ISBN 0-8155-1148-5 (1988)

697 pages

Printed and bound by CPI Group (UK) Ltd, Croydon, CR0 4YY

14/10/2024

01773862-0001